COLLECTING

Fossils

by Alan Major

Illustrations by Barbara Prescott
photographs by Tom Scott

ST. MARTIN'S PRESS, INC.
New York

St. Martin's Press, Inc., 175 Fifth Avenue, New York, N.Y. 10010
Printed in Great Britain
Library of Congress Catalog Card Number: 75-7729
First published in the United States of America 1975

Contents

Acknowledgements

The publishers acknowledge the assistance of Mr. William Baird, Dept. of Geology of The Royal Scottish Museum, who made available the exhibits for the colour photographs, which were photographed by Tom Scott (Edinburgh).
The author acknowledges with sincere thanks the assistance of Christine Hartley, Victoria Island, B.C., Canada, regarding North American fossils; Mr J. C. Thackray at the Geological Museum, London, for his generous assistance with the British Geological Time-Scale and Stratigraphical Column; and the co-operation of the book's illustrator, Barbara Prescott, in her preparation of the illustrations which made the task of producing the book so much easier.

LIST OF COLOUR PLATES

Dedicated to my wife
Jean
for her patience in
checking the manuscript

Author's Introduction

Fossil study has a reputation for being a rather dull subject. Fossils are, admittedly, old, dead, inanimate objects which at first might be thought to be uninteresting, but they are, in fact, evidence from millions of years past when life was very different on earth. When the background story of fossils is known it is soon realised how fascinating they are and without them much of our knowledge of the evolution of present day forms of life would still be unknown.

The collecting of fossils can also be an absorbing hobby and can be as serious as the collector desires to make it. He or she can specialise in one aspect, for example Plant Fossils, although by doing so they may restrict their chances of obtaining fossils and by taking an interest in a wider range of fossils more periods and thus more fossils sites are available to them.

The British Isles and North America are very fortunate in that they have exposures of rocks from almost all the geological periods. This means that wherever you live fossiliferous rock outcrops and localities are within range, perhaps even in your own back garden. There is also no restriction on when you can go fossil hunting. At any time of the year and in any weather, if you want to, fossils can be sought.

Perhaps most fossil enthusiasts begin by digging up a sea-urchin in their garden, or find a belemnite on the seashore beneath a cliff face, or similar, and their curiosity is aroused. If they succeed in identifying it and learn some of the example's history then more are collected and so on. My book is intended for those people who casually find a specimen and need a book to identify it, with information on how to begin fossil collecting as a hobby. Secondly, the book is intended as a reference work, both in the British Isles and North America, for those collectors who need to know more about the geological periods and the fossils they contain in each other's country. This can be especially valuable to collectors who correspond with collectors in other areas and swap fossils by mail.

I have endeavoured to write the text and descriptions as simply as possible, but the majority of fossils have scientific names and so use of these was unavoidable.

ALAN MAJOR

1 How to Collect

CHAPTER I

The collector of fossils does not require costly and intricate equipment to obtain and preserve specimens. Several of the items necessary are quite ordinary and easily obtainable, but as he or she progresses in experience, if he or she so desires, the collector can add more sophisticated equipment.

It is very important to have a good hammer or hammers. The temptation for the amateur is to buy a carpenter's or bricklayer's hammer. These would be adequate on soft rock, but there is always the danger that while striking a chisel on very hard rock the soft steel hammer head itself may split or shatter, with the possibility of resulting injury to the collector. So I advise the purchase of a geologist's hammer of special hardened steel, which is more suitable for all purposes and for safety's sake. This usually has a flat end for striking, the other end being either bevelled for use as a chisel to lever away the material containing fossils, or, alternatively, it may taper to a point. The balance of the hammer is also important in use and so the shaft must be of the correct length. For a hammer of 2 to 2½ lbs. weight the shaft should be 18 inches or longer. If obtained from a reputable geological supplies company the hammer will be designed with the right length of shaft. All-steel hammers are suitable for use but a truly-weighted hammer with head correctly attached to a wood shaft has a 'feel' about it that gives confidence. In dry climates all-steel hammers are essential if shrinkage of the shaft in the head is a problem, but not one regularly met with in the British climatic conditions! A range of various-weight hammers should be obtained for use in different geological occurrences. For example, a 7 lbs. weight sledge-hammer used on chalk would be literally following the moral of the saying 'to use a sledge-hammer to crack a nut'. A 2 lbs. or 2½ lbs. geological hammer is a good all-round weighted implement, but the serious collector should also have lighter examples, of ½ lb. and less (the type of hammer once used by confectioners to break up slab toffee is ideal for delicate work), a 1 lb. hammer, also a 4 lb. is useful, but for the very hard rocks larger, heavier hammers, up to 14 lbs. are necessary. For use on chalk a tiler's hammer with a pointed end, or a plasterer's hammer, could be used as a makeshift alternative.

The next item is a range of various-sized cold chisels for use with the hammer to remove the fossil from the rock or other material. A general purpose size is one with a sharp-edged, half-inch blade. For the finer removal work a quarter-inch blade is handy and sometimes vital, while for the extraction of bigger rock portions containing the fossil, or to split the material, larger, longer cold chisels, up to 18 inches in length and 2 inches or so in width should be among the collector's set of chisels. These can also be obtained from reputable geological supplies companies or ironmongers may have specially hardened cold chisels in stock.

To prise a section of material apart a crowbar is useful. If the material to be worked is at a depth a pick will also be necessary. A shorter-handled, lighter-weight pick comes in handy to remove the material surrounding a fossil. A mattock, a type of axe, can be used with care for this purpose. For extracting fossils from clay or soft chalk a carpenter's blunt chisel with a wood handle could be used as an alternative. A strong penknife can be very useful for removing fossils from some of the softer rocks, mudstone, chalk, etc. The penknife can also be used to cut away some of the material around the fossil if extracted as a lump.

For the softer materials flat-bladed trowels, normally for cement pointing, can be used for part of the work of extracting the fossils, the task to be finished with a strong pocket knife.

Where fossils are contained in sand a small meshed sieve is useful, with a shovel or trowel to lift the sand into it. By using these some of the smaller fossils which might otherwise be missed can be obtained. The sieve could be used for washing fossils obtained from other materials if there is a supply of water in the vicinity but do not roll the fossil specimens vigorously around in the sieve as the impact of these upon the mesh or sides may damage frail specimens.

A soft-bristled brush, about 1 inch or so wide, is also handy to have with you to remove material obscuring the larger fossils, as extraction is taking place. For the tinier, fragile fossil specimens, where the surface can so easily be worn away, a child's soft paint-brush can be used for the same purpose. But do not brush the fossil entirely clean of powdery or other material as this may under close scrutiny, reveal details not at first apparent to the naked eye. It is equally valuable to be able to distinguish from its remains in which type of deposit the fossil was found. For this reason many museum specimens are not scrubbed clean and are also embedded in a fraction of the deposit material.

A pocket-lens magnifier is also useful for fossil examination on the site, particularly for the smaller specimens, to identify them and to see if they are damaged. To examine fossil remains, such as pollen, grains and spores, in peat, a microscope with good magnification is necessary.

After removal it is essential to wrap each fossil separately in a piece of newspaper or other paper, so some sheets should always be taken, to protect the fossils from damage while being carried. A toilet roll is a convenient way of having a supply of small sheets of paper available to make into 'twists' for the tinier fossils. As an extra precaution these latter can also be put in polythene bags, which help to prevent the paper unrolling and releasing the fossil, while being transported. The very frail or small fossils should be placed in small boxes or tins lined with soft paper wadding or similar material, but not cotton wool as this is sometimes difficult to remove from fossil parts. Spare paper should be put in the spaces between the specimens to pack them securely but not too tightly and to prevent movement.

The fossil collector should also have two pencils or biro-type pens, one to use and one as a spare, a notebook and some strips of white paper. As each specimen is removed from the material and put in the newspaper or container, a number, letter or identifying code, should be written on a paper strip and placed with the specimen. The same number, letter or code should be written in the notebook and alongside it the relevant details required to identify and label the specimen – locality, position on the map of the area, level of the site, material in which found, date, and, if known, brief identification of the fossil. There is nothing worse than arriving home with several fossils, unwrapping and placing them on a table, then racking one's brains to try and remember the foot of cliff-faces and in rock exposures. On the seashore with fossil-bearing cliffs, rocks or sand, the collector may be able to simply pick up fossil fragments, which in the beginner will stir his or her enthusiasm to greater efforts. This is also a clue to the experienced fossil collector that there are possibly better fossils awaiting discovery in the cliffs and rocks at a more inaccessible, higher level or among the sand and clay deposits.

New, man-made exposures, such as road and railway cuttings, trenches for pipe-laying, removal of material in quarries and similar excavations, are always worth examination in fossiliferous localities. So, too, are the waste heaps from coal mines. Here may be found heavy ironstone nodules that at first sight may not look interesting, but they frequently enclose various fossils, of fish, arthropods, insects, etc. Sometimes they cannot be broken open without the hammer's impact shattering them, but this can be overcome by leaving them in the open for several months so that exposure to frost, ice and rain can play its part. Then when dealt with, they usually break open cleanly and the fossil within can be obtained. A quicker method, if you cannot wait for the winter to satisfy your curiosity of what is contained in the nodule, is to soak the nodule in water, then put it in a refrigerator for several days. On removal from the refrigerator drop it into boiling water, whereupon it should split open along its natural plane of weakness without ruining the fossil therein. It is wise to protect the eyes while performing this task just in case splinters do fly off the nodule's exterior at its sudden change in temperature.

River banks may collapse or be scoured away to reveal fossils. If such places as railway cuttings, quarry faces or mine dumps have been left untouched for years they are equally good sites to examine. Erosion of the material by the weather may have left many of the fossils on or projecting from the surface. Visits to the tips near a coal mine to examine the unwanted shale are usually fruitful for a wide range of fossilized vegetation.

Of course, if visiting a quarry or other exposure situated on private land the thoughtful fossil collector will seek permission first, either verbally but preferably in writing, from the person in charge of the workings or, better still, from the land-owner. Provided the site is safe enough most owners will grant permis-

There are two ways of dealing with them. One is to cut away the surrounding area as a large lump, a sufficient distance from the fossil, so that when it is lifted away the risk of the fossil's disintegrating is lessened. Then it can be taken home to be dealt with, by hardening it, before reducing the lump to a more appropriate size as a background to the fossil. Alternatively, the fossil can be hardened on the site to reduce further the risk of damage. Exceptions are fossil leaves and plant remains as these have a tendency to curl over in distortion if hardened. One hardening material is amyl acetate and acetone in equal parts mixed with Rawlplug Durofix. It should be brushed on lightly or can be poured on and then allowed to harden and create a thick glossy surface coating over the fossil. This liquid is inflammable, however, so should be kept away from naked flame, even a lit cigarette, while the task is being done; otherwise a fossil which has lain preserved in its matrix for millions of years could be destroyed in a few seconds. The liquid solution called 'Knotting', obtainable in oil and paint shops, can also be used. If it is too thick it can be diluted with methylated spirit but needs to be of a consistency to penetrate the frail fossil for complete hardening and protection. The disadvantage of this solution, however, is that it stains the fossil a yellowish-brown colour and much of the fossil's beauty may be lost.

Those fossils which are composed mainly of pyrites have a tendency to powder after a few months or years of exposure to air following removal from the material in which they were contained. The hardener can be used on small specimens of this type, too. Frail fossils and bones can be prevented from deterioration and cracking by coating with a layer of shellac or Alvar. Another way to preserve and protect frail fossils is to embed them in a cold hardening plastic such as Isopon. This can also be used as a means of display instead of in cases or boxes. However, the larger as well as the smaller specimens, can be preserved more simply by brushing them with hot, pure, clear candle wax. Only a light coating is needed. Do not coat too thickly or the wax may break away on cooling and hardening and take some of the fossil surface with it. If too much is applied accidentally this can possibly be carefully scraped off with a penknife or can be removed by equally carefully wiping with carbon tetrachloride sprinkled on a piece of cloth. A third method against pyrite decay is to soak the fossil in I.C.I. 'Savlon' Solution.

Do not under any circumstances attempt to dry out fossil specimens that were damp when found in the deposit, or made wet to be cleaned, by placing them over or near to heat. They should be allowed to dry very slowly, naturally. Otherwise powdering, crazing, splitting or other deterioration can be caused.

After removal from the surrounding rock or material the fossil or revealed side will require cleaning. The hard fossils can be washed using e.g., a soft toothbrush, in warm water, until the unwanted material obscuring the fossil is dissolved or scaled away. Sometimes, holding the fossil under a running cold water tap is all that is needed. A fossil removed from chalk or limestone may need to be soaked for a while in warm water. Then a stiffer, but not too stiff, brush may be used to remove the chalk or limestone, but do not scrub the fossils too vigorously or the surface may be damaged or removed. Hard chalk around a fossil can be treated by placing the fossil and chalk in a dilute solution of hydrochloric or acetic acid. Care must be taken with this treatment and rubber gloves worn when handling the fossils as the acid will burn the skin. Use of a toothbrush or knife will remove the chalk, but a disadvantage of this treatment is that the acid may also penetrate and damage the fossil.

Very frail fossils that look more like imprints in the rock must not be brushed at all. A needle fixed into a wood handle is a useful tool to scrape out and remove any adhering chalk or material in the grooves or ornamentation of the fossil.

It is not impossible to repair a shattered fossil or one with a portion broken from it. If the fossil is broken cleanly there is a chance it will not be too visibly damaged. Where there are numerous pieces badly splintered it will be very obvious that it has been repaired. When the damage is extensive the collector must decide whether the value of the fossil is worth the repair. Personally I feel it is, because the damaged fossil can always be kept until another specimen is obtained to replace it. If a replacement is not found then the damaged specimen fills the gap there would otherwise be in the collection. There are several suitable adhesives for this purpose. One is 'Araldite', rather slow to set completely; others are 'Uhu' and 'Durofix', which are much faster; a third group of suitable adhesives being the 'P.V.A.' glues for the large fossils.

After cleaning and identifying the fossils they must be adequately housed. How this is achieved depends on the circumstances of each collector, both financially and as to available space. Some of the store cabinets used for other forms of natural history collecting can be adapted for use in accommodating and exhibiting fossils. Ideal is the glass-topped table-type cabinet but the cost of new examples may be too prohibitive for the beginner-collector. By attending sales or furniture auctions it may be possible to buy one reasonably cheaply secondhand or, failing a proper cabinet, there may be an item of furniture in a secondhand shop that can be altered for showing fossils. Glass-fronted bookcases or display cabinets for china can be used for this purpose. A chest of drawers, tallboy, or old wooden filing cabinet can sometimes be converted successfully, although the specimens are not open to view.

If at all possible it is preferable to have drawers of different depth so the small examples can be accommodated in the shallow drawers, the bigger fossil specimens in the deeper drawers. Sometimes, as an alternative, it is possible to purchase an old-fashioned miniature chest of drawers, up to 2 feet in height, 18 inches or so wide, which stood often on another piece of furniture, such as a sideboard or on a table, and was originally used for small items, trinkets, letters, etc. These are ideal for storing the small fossil specimens.

When adapting a chest of drawers, filing cabinet or similar, some of the drawers should be divided into compartments. Lengths of plywood, fitting into slots at each end to stay in position, are suitable. Some of the larger, deeper drawers should be kept so that these can house the large blocks containing fossils.

A collector who is a reasonably skilful carpenter, or who has a friend or relative who is, can adapt wooden boxes for use to display fossils. Lengths of wood can be glued around the inside rim of the box, in between which a sheet of glass or clear plastic is then inserted to slide in and out as a lid.

A shallow wooden box can be made into a wall case to display the smaller fossils. A sheet of thick white card can be used as the background, pinned to the bottom of the box. To this are fixed the fossils by using adhesive on their bases. As extra support two or more pins, with their heads removed if preferred, or fine insect-setting pins, can be pushed in position underneath each fossil to take the weight of the fossil as it hangs upright. Over the box fix a sheet of glass or clear plastic so the specimens can be seen. It is not advisable to hang the fossil display in a position where the sun regularly shines on it, as this may affect the surfaces of some of the fossil examples. Nor should the boxed display be too close above central heating radiators.

If there is space available in a spare room where it is intended to keep the collection another method is to erect strong shelves on which cardboard trays or boxes containing the fossils can be stacked. Alternatively, a wooden rack with several rows of shelves to hold boxes of fossils can be constructed either from wood or Dexion. Strong cardboard boxes of varying sizes and depths are easily obtainable. They can be divided into compartments by using lengths of spare card glued into position or each fossil can be separately put into a cardboard box about its own size, these smaller boxes with contents being put in a large box for stacking on the shelf. It is not advisable after the effort of carefully obtaining them to heap several fossils upon each other in boxes or drawers. If examined frequently the sliding in and out of the drawers, or lifting and replacing of the boxes, will cause the shells to jolt against each other so they may be scratched, worn or even splintered. The drawers and boxes should be lined with a soft material, such as foam rubber, felt, or cotton wool, to prevent this unnecessary damage. For the very small fossil examples, matchboxes are ideal. A number of these can be made into a miniature chest of drawers simply by glueing some of the matchboxes above and alongside each other. To improve the exterior it can be covered with white or coloured gummed paper, a suitable piece of spare wallpaper, or adhesive plastic sheeting. The trays should also be lined with thin felt, foam rubber or cotton wool to protect their contents. Where small fossil examples of related larger specimens are obtained, such as the sometimes tiny, sometimes massive ammonites, the small examples can be housed in matchboxes or similar sized boxes and kept in the same drawers as the large specimens. Extremely tiny fossils or fossil fragments can be placed in glass tubes. These are obtainable in various sizes from science equipment suppliers. Smaller tubes can be kept after tablets or pills from the chemist have been used. Such tubes can easily be housed in considerable numbers inside a separate cardboard box. If needed on display they can be mounted on test-tube racks which are also obtainable from science equipment suppliers, but are also easily made from several pieces of wood with appropriate-sized holes bored into one piece on the top, in which the tubes will stand upright. A piece of cotton wool should be placed in the bottom of the tube in which the fossil specimen will lie and a plug of cotton wool placed above it in sufficient quantity to prevent the fossil moving up and down the tube's length so causing the fossil to become worn or otherwise damaged. Each tube should also be sealed with a cork.

Lastly, however the collection is stored, every specimen should be labelled with as complete a set of detailed information concerning it as possible. There is nothing more frustrating or near-useless as fossil specimens without labels. And, no doubt, the value of a collection is increased when concise details are stated. For this reason the fossil collector should carry a notebook and spare biros or pencils, as I have previously advised in this Chapter, in order that at the same time as placing a letter, number or code on the paper strip to accompany each fossil as it is wrapped, he or she also puts the same letter, number or code in the notebook with the details surrounding the circumstances of the fossil's discovery. If the collector does not do this and attempts to write the labels at a later date there will be difficulty in remembering accurately all the facts and it may even entail another visit to the site to check on some of the details. This could be inconvenient if the collector lives a long distance from the fossil site.

On confirming the identity of the fossil the English name should be written on the label, in capital letters preferably, below it the scientific name should be written in capital and small letters, followed in clear writing by other relevant details, such as where found, material surrounding fossil, age of fossil if known, date of discovery, finder's name. Labelling can be taken further if required by appending other details about the fossil if they are known, such as the type of life it lived, i.e., whether swimming

1. Mollusc becomes fossil

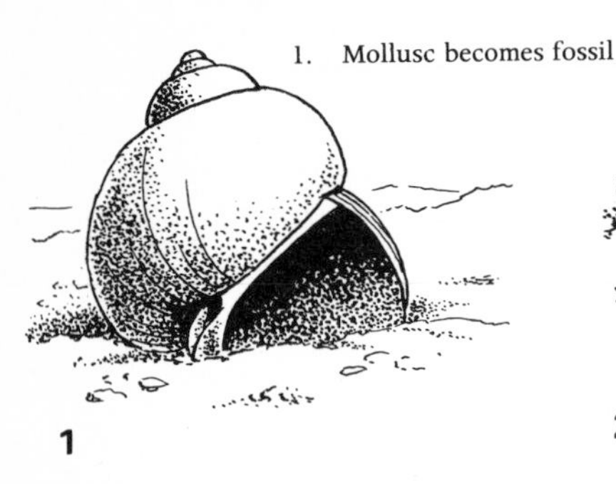

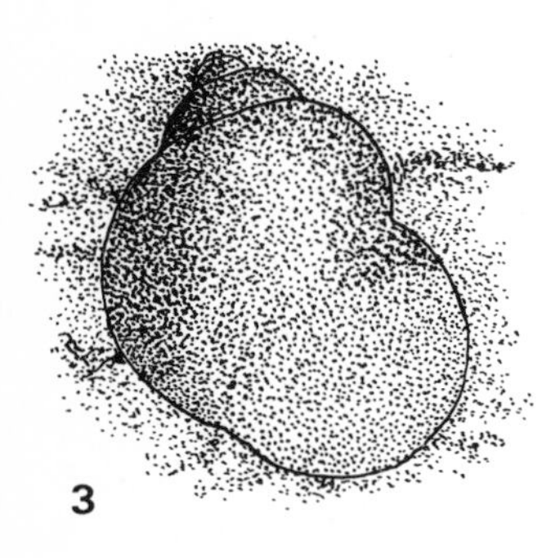

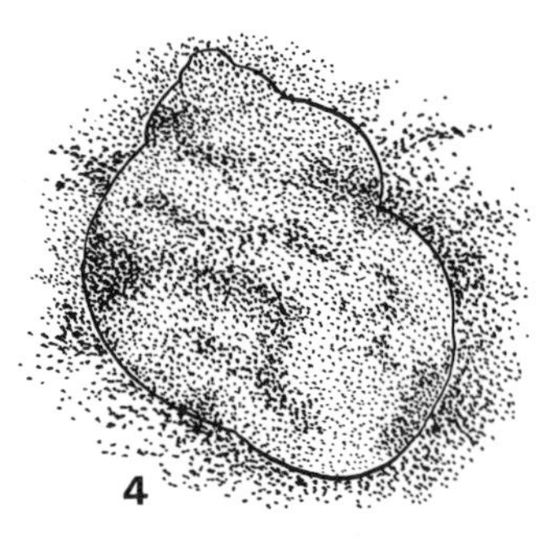

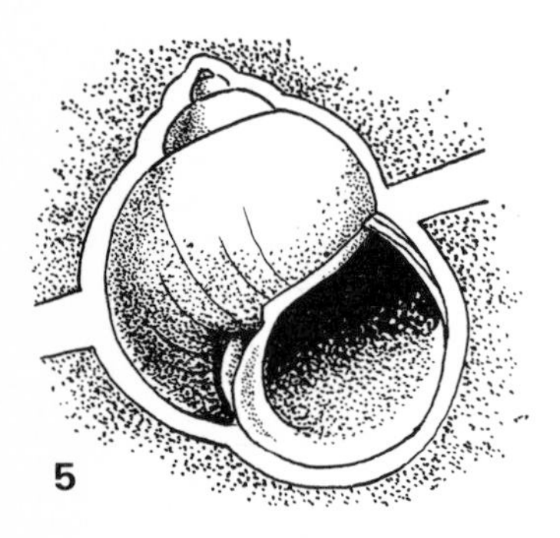

1. Death of the mollusc and shell is covered by sand.
2. The sand is hardened to form rock, but the mollusc shell is not filled in.
3. The shell then dissolves, so the empty cavity is the mould of the shell.
4. Dissolved minerals fill the mould to create a cast of the shell.
5. The mould and resulting cast in it are fossils, both facsimiles of the original mollusc shell.

sea creature, crawling on sea bottom, or land surface, plant or flesh eater, if vegetation whether a swamp inhabitant, etc. The writing should be in Indian or waterproof ink and a fine-pointed pen should be used upon thick paper or card. When completed the label can be placed or pinned alongside the specimen in the compartment containing it in the cabinet, drawer or box. It should be pinned in position under the example in a wall case, not glued, as the specimen may later be changed for some reason. A new label would be required if glued, but is easily changed if only pinned in position. Thinner paper can be used if the label is to be inserted inside a glass tube. Also put a label on the larger cardboard boxes listing their contents. This prevents having to search each box to find the fossils required.

It is also valuable to keep a log-book in which the same letter, number, or code is given, then alongside this the same details can be copied. An advantage of this is that while only brief details can be written on the display labels, in the log-book these can be described more fully, concerning the actual site of discovery, if and what other fossils were found in the close vicinity, their ages if of a different era and so on.

As an alternative, a collector may prefer to use a card index system with a separate card for each fossil specimen. On this card is stated the same details with relevant figure, number or code as for the log-book method of recording details. These cards can be filed in alphabetical order, but preferable is the method of placing them in Eras which are themselves divided into the various Periods or Systems, the latter again being divided into categories, Fish, Reptiles, Shells, etc. The card index system does have the advantage that a mass of information can be immediately available for study, but is generally immobile. A log-book's disadvantage is that it can be lost or destroyed and with it perhaps years of note-taking, but it is portable and can be taken wherever required. But each collector will decide on and develop the system which he finds easiest and most practical for his own purposes of quick reference.

Similarly a collector may have his own ideas on how he can best store and preserve his fossil collection, which suits his own particular requirements and if this prevents the fossils from being damaged or from decaying while housed and keeps them free of dust and in good condition while available for inspection, then it will be satisfactory.

2 Fossils and their origin

A fossil is the remains of an animal or vegetable organism, found petrified and naturally embedded in stratified rock. This definition also includes the casts or traces of such organisms, for example, the tunnels or trails left by a worm's movements, or a footprint. The word fossil, from the Latin 'fossilis', means 'that which may be dug up'. But this is not entirely accurate, because it concerned everything dug up or discovered in the earth, the crystals, minerals, coloured and unusually shaped stones, etc., as well as the fossils.

What we now know to be fossils were then called 'figured stones'. How they came to be buried was explained by early investigators simply that the world was overcome by a series of 'Great Floods' or 'Deluges', which drowned and buried all living things in their path, as well as stones and rocks. Writing in volume 3 of his 'Spectacle of Nature', published in France in 1733, a work which was very accurate in some aspects for the period, Antoine Noel, a notable French naturalist in the early 18th century, still referred in a footnote to 'whatever is or may be dug out of the Earth we call a fossil'. He does, despite this, describe how 'pieces of Wood, Shells, and Other Substances found petrified Under Ground still retain their natural Figure and Linearments by Insertion . . . water performing all the offices of collecting, conveying and associating all the different Particles of Matter which afterwards condense into Aggregates of a hard durable Nature . . . Water may easily introduce various Kinds of Matter into the Pores of a Body and thereby bring it to a State of Petrifaction' (now known to be a fossilization process). Noel then continues in detail to describe various fossils and surmises fairly accurately how the 'petrifications' came to be buried in the 'strata of Matter extended under the Earth'.

A big step forward was that of William Smith (1769–1838) who studied the rocks revealed during the excavations for the

Kennet-Avon canal and the fossils the rocks contained. He laid the foundations of stratigraphical geology by showing how 'strata can be identified by their contained fossils', enabling rocks to be dated because the fossil origins they contain only existed in a particular Period of geological time. Various naturalists and other interested men, whose work is related in detail in some of the books listed in the bibliography, investigated and have gradually uncovered the story of how fossils were created and laid the foundations of Palaeontology, the study of fossils, from the Greek for 'discourse' (logos) on 'ancient' (palaios) 'beings' (onta).

To understand how fossils were created it is necessary to know how the rocks containing them were formed. There are three main types of rock–Igneous, meaning fire, made from molten volcanic lava; Metamorphic, meaning changes, rock which has been changed in chemical or physical condition, by heat, or pressure; and Sedimentary, formed from sediment flowing via rivers from the land into the sea. It is the Sedimentary rocks which contain fossils.

By the gradual depositing of sediments eroded from already existing rocks over a period of millions of years the Sedimentary rocks were laid on the sea bed, but also in estuaries, lakes, marshes and swamps. Equally slowly over millions of years the sediments hardened into limestone, chalk, shale and other soft or hard rocks. At some period in their history these rocks, sometimes in huge masses, were forced upwards and in this process were raised above the level of the sea and formed new land.

Prehistoric plants and animals that lived on the land were unlikely to become fossilized. As soon as they died their bodies were either consumed by scavenging creatures or decayed by the action of the weather, so that nothing remained which could become fossilized.

But there are several exceptions, although they are not fossils in the true sense. Of these, Insects, have been preserved through being covered by a quantity of hardened resin from trees. This resin, called amber, encloses the victim and prevents its decay. Pine forests in the Baltic region of Europe are a good source of these. A second exception is where the living or dead animal has fallen into a bog or pit of a preserving substance. Peat is one example, although in all cases only the bones remain, the soft parts having disappeared through decay. Natural tar pits, such as those of Rancho La Brea, near Los Angeles, California, U.S.A., and southern Poland have given up their prehistoric victims in a good state of preservation, including the soft parts. The skins of dinosaurs have been found mummified by the dry atmosphere of Patagonia, South America. Alternatively, woolly mammoths have been discovered in Alaska and northern Siberia because their bodies were preserved in the frozen mud.

For true fossilization to occur the corpse must be entirely enclosed and buried soon after death. (Fig. 1) On the earth for land-living or winged organisms this did not happen for the reasons stated, but those organisms which lived upon the shore of a lake or the sea, in the water, or upon the bed of the sea or lake, there was a greater possibility of fossilization. When these organisms died their bodies descended through the water and lay on the sand, mud or gravel. Their soft parts rapidly decayed, but the harder parts, shells, bones, teeth and such like survived. These quickly sank into or were covered by a layer of the sedimentary mud, sand or gravel and so were protected from further destruction by the chemical action of the water and still living boring organisms. Thus they became fossils.

The most common fossils found are mollusc shells, sea-urchins' tests, the teeth and bones of birds and mammals, the leaves, branches and trunks of trees, and vegetation leaves. The impression or trace made on fine-grained mud by the soft body of a jellyfish, before it decayed, and preserved is classed as a fossil. Borings, trails and footprints made by organisms in the originally soft mud or sand were preserved underneath when the mud or sand hardened enough to prevent the imprint or boring from being destroyed when covered by a layer of sediment.

The fossils of some organisms have remained virtually unchanged since they were occupied by their living owners. Others have undergone considerable change in their composition.

The hard parts or framework of the organisms are usually of the inorganic mineral calcium carbonate, calcium phosphate, silica or chitin. Shells which comprised calcium carbonate in an unstable form called 'aragonite' are usually converted, after burial, to a different form called crystalline calcite. This occurs when water percolates through the material or sediment enclosing the organism. In the case of shells they have a fibrous composition and are slightly porous. The water dissolves the original fibres of the shell and replaces them by solutions in the water filling in the shell's structure so that it is preserved, but instead of the original shell there is a perfect, solid facsimile. In some cases the water substitutes the shell composition with a mineral, if the latter has been extracted by the water from rocks, such as copper carbonate (malachite), oxide of iron or iron sulphide (pyrites).

Sea-urchins are often found inside when the massive flints that form layers in the chalk are split open. The same process that fossilized the shells occurs in this case. The sea-urchin has a thin, brittle, rigid limy shell or test, an inner skeleton of tightly-fitting plates, with small holes or pores from which the tube-like feet and moveable spines project. After death and enclosure the soft internal parts of the body inside the test decayed and were replaced by silica which percolated into the inside of the test via the holes and formed the internal mould, after which water eventually dissolved the limy test. The broken-off spines are also found fossilized in the chalk.

In other instances the seeping of the water, especially through porous rocks, dissolves the shell or skeletal element entirely and all that remains is an empty cavity. Such cavities are valuable for collection because they still bear the impression of the formation, ornamentation and other structures of inner and outer surfaces of the shell or other organism. These cavities are known as external or internal moulds of the original, depending on the impression on the rock being either the original's external or internal side. The shape and pattern of the original occupant can be obtained as a cast by placing a suitable material, such as Plasticine or 'Vinagel' in the cavity.

In suitable circumstances tree trunks, branches and roots have also been fossilized. The cell structure of the tree–soft, organic material–is infiltrated grain upon grain by silica, from siliceous, hot, spring waters, so that eventually the silica replaces the original cell structure. But the latter is faithfully reproduced and preserved in minute detail. This is known as 'petrifaction' and there are several well-known examples. At Lulworth Cove, Dorset, there is a 'Fossil Forest' comprising the silicified stumps of coniferous trees sheathed in tufa. In Arizona, U.S.A., and Patagonia, South America, entire tree trunks have been 'petrified' by silica.

Similarly, when calcareous water ran from a 'petrifying spring' or other source into a peat bog the lime impregnated the remains of trees, plants and other vegetation, then crystalized as calcite and this sealed the tissues of the remains and preserved them. Sometimes these formed round boulders known as 'Coal Balls', found in coal seams, and microscopic examination of these has revealed information about the Carboniferous plants.

Among peat accumulations are many fossil remains, chiefly fragments of tree trunks, branches, plant stems, also leaves, seeds, tiny pollen grains and spores.

Coal itself is fossilized peat. On the fringes of coastal lagoons and swamps, coal forests, with ferns, clubmosses and other types of vegetation, grew luxuriantly. What remained of this vegetable matter when it died fell and sank into the swamp to accumulate, or was buried by river-borne mud or sand sediment, or land subsidence. It rotted and slowly underwent (slow by geological time standards) chemical alterations, carbonisation and the consolidation of coal. In coal and shale many fine examples of plant fossils can be found. These are in the form of impressions with the outer shape and surface markings, as carbonaceous films, or as petrifications.

What is most fascinating about fossils is that they are evidence of forms of life which existed millions of years ago and which in some instances have vanished forever.

These forms of life began as very primitive organisms, sponges, algae, worms, etc., from which developed the larger groups of plants and animals. Palaentologists have been able to trace the way in which these animals and plants developed to their present-day forms. But a large number of forms of life were unable to adapt to changing conditions and eventually became extinct. One example is the ammonites, some types of which are commonly found. They died out approximately 70 million years ago. Today the closest living example resembling the ammonite is the pearly nautilus, which occurs in the warm

Plate I Fossil Wood, from the Petrified Forest, Arizona, USA.

Plate II Antler of Irish Giant Deer, *Megaceros giganteus* (Blamenbach). Pleistocene, Ireland.

waters of the Pacific Ocean.

As well as being examples of organisms that lived in the Earth's infancy fossils have an important use in assisting geologists to co-relate rocks occurring in different parts of a county or counties (Fig. 2). By comparing the fossils they contain, a relationship to each other can be established between rocks and strata. It is also possible to establish by studying fossil locations, that various now separate land masses were originally linked. The fossils are typical of the conditions prevailing at a certain time or age. Evolutionary changes occurred in the organisms that eventually became fossils, and so by studying the fossils contained in rocks or strata it is possible to date the rock or stratum and to determine its position in relation to others.

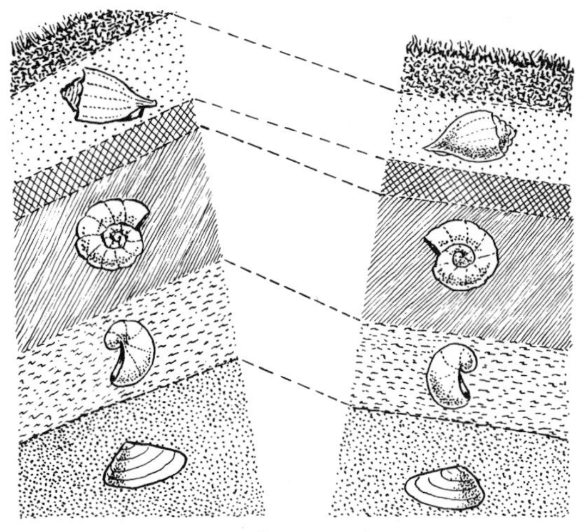

2. Co-relation of strata

Geological history has been arranged in a 'time scale' and Stratigraphical Column. The latter indicates the order of sedimentary rock formations in the sequence in which they were deposited. The first division is into Epochs or Eras–the Pre-Cambrian or Pre-Phanerozoic, the Lower and Upper Palaeozoic, Mesozoic, Tertiary or Cainozoic, Quarternary. These are divided into Periods or Systems. The System refers to the rocks in the Stratigraphical Column and the Period indicates the time when the rocks were laid down, thus when referring to time involved it can be described as, for example, the Carboniferous or Silurian Period, but when referring to the rocks it is given as the Carboniferous or Silurian System. The Systems are also divided into Series or Formations of which there may be several. For example, Lias and Inferior Oolite are two of the Series or Formations in the Jurassic System of rocks. By knowing the Stratigraphical Column it is much easier to appreciate the time involved and simpler to date the fossils that are discovered when prospecting in the Series or Formations and thus know where they fit in geological history. I have also listed the Column separately, page 58 for quicker reference, arranged in sequence from the earliest Period and progressing to the most Recent Period and System.

Here it will be of interest to explain the meaning, and how, the names used in the 'time-scale' and Stratigraphical Column were derived. Firstly the Eras or Epochs, which are named to correspond with their fossil content and order of deposition: Palaeozoic, meaning 'ancient life'; Mesozoic, meaning 'middle life'; and Cainozoic (Cenozoic) 'recent life'.

The names of the Periods or Systems refer either to geological sites or features, inhabitants of these areas, or the age of the Systems. Cambrian was so-named after Cambria, the Roman name for Wales; Ordovician after the Ordovices, a Celtic tribe which lived on the Welsh borderlands where this rock is frequent; Silurian after the Silures, a Celtic tribe as the foregoing; Devonian named after the county of Devon, England, where this System was first identified; Carboniferous (British time-scale) named from the existence of carbon- or coal-bearing rock seams in the British Isles; Mississippian (North America time-scale) named after the Mississippi River limestone outcrops; Pennsylvanian (North America time-scale) named after the coal-bearing rock seams in Pennsylvania; Permian named after the province of Permu, Russia, where the British geologist pioneer Murchison worked for the Russians; Triassic named after typical three-fold division, sand-limestone-marl, of beds in an area of northern Germany; Jurassic named after the Jura Mountains, on the French-Swiss border; Cretaceous derived from 'creta' the Latin word for chalk; Palaeocene from the Greek 'ancient-recent', meaning 'the earliest of recent forms of life'; Eocene meaning

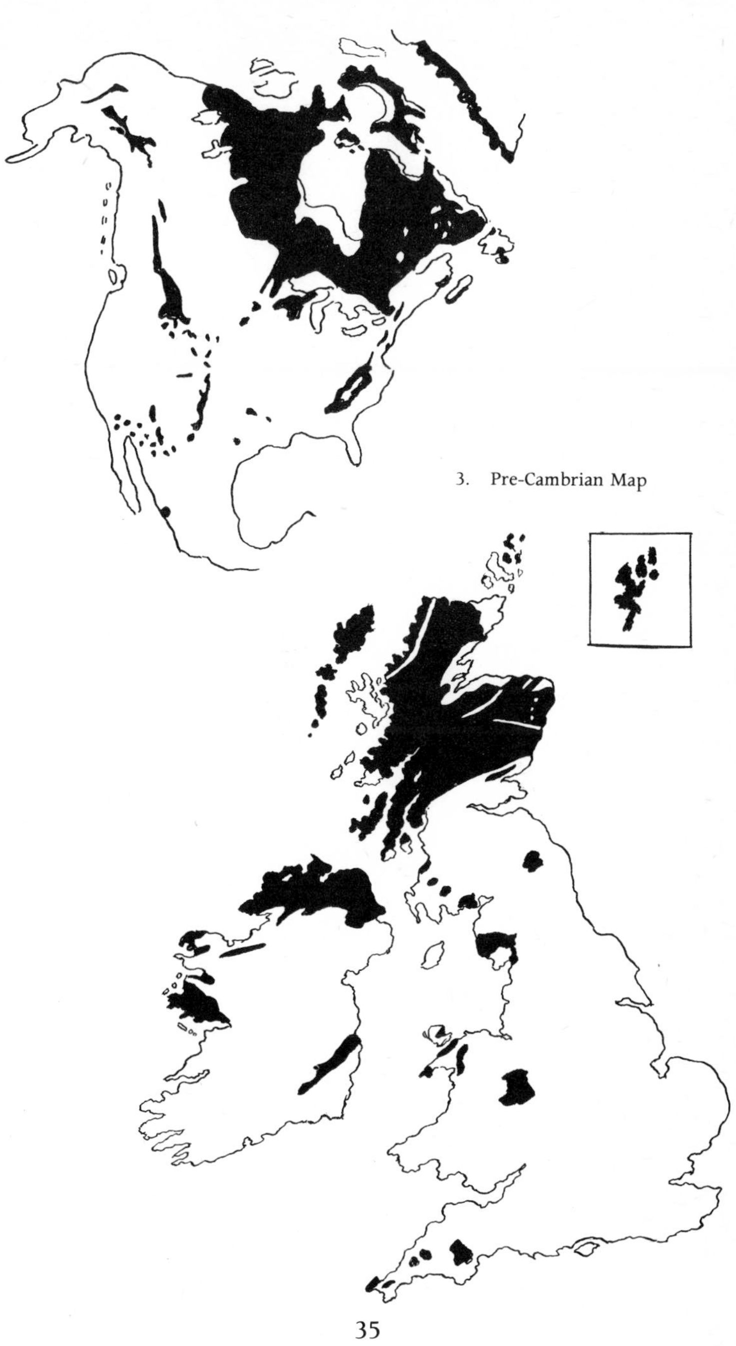

3. Pre-Cambrian Map

'the dawn of recent life'; Oligocene meaning 'few or little recent forms of life'; Miocene meaning 'less recent forms of life'; Pliocene meaning 'more recent life'; Pleistocene meaning 'most recent life'; Holocene meaning 'wholly recent life'.

THE PRE-CAMBRIAN PERIOD (Fig. 3) is the Period of time that passed before the first clear fossil-bearing rocks were deposited in the Cambrian Period. It totalled approximately 4,500 million years and during this time began the origin of life and the development of primitive living things. Even so, fossils from this Period are extremely rare for several reasons. So primitive was their construction and so soft-bodied were they, there was virtually nothing to be fossilized. Those that did become fossils in the Pre-Cambrian sediments were later destroyed by changes and movements during the metamorphism or deformation of these sediment rocks. In some instances where the Pre-Cambrian rock is unchanged from the time it was laid down it is of a rock type, such as Red Sandstone, in which fossils are not usually enclosed. There are Pre-Cambrian schists and gneisses in areas of England, Ireland, Wales and Scotland, and Pre-Cambrian rocks in areas of Australia, South Africa, Rhodesia, Canada (Ontario), and the U.S.A. (Minnesota). Pre-Cambrian fossils of worms have been found recently in the Ediacara Hills, South Australia. In Canada fossils of a fragile, tiny organism, called *Eozoon*, which may have been animal or plant, have been found in rocks about 2,600 million years old. In Montana, U.S.A., a lime-secreting algae, called *Collenia*, has been discovered in Pre-Cambrian rock, while from the Grand Canyon, Arizona, U.S.A., a jellyfish fossil has been recovered.

THE PALAEOZOIC ERA

THE CAMBRIAN PERIOD (Fig. 4) rocks are the oldest fossil-bearing rocks and prove there was at that time a variety of organisms in being. In the British Isles they are chiefly Sandstone and Mudstone occurring in east Ireland, northern Scotland, Isle of Man, North and South Wales, and Shropshire, England. In the Lower Cambrian System there were common though primitive sea-living Algae, Arthropods, Brachiopods, Trilobites, Sponges, Coelenterates, Echinoderms, Worms, Molluscs. In the Middle

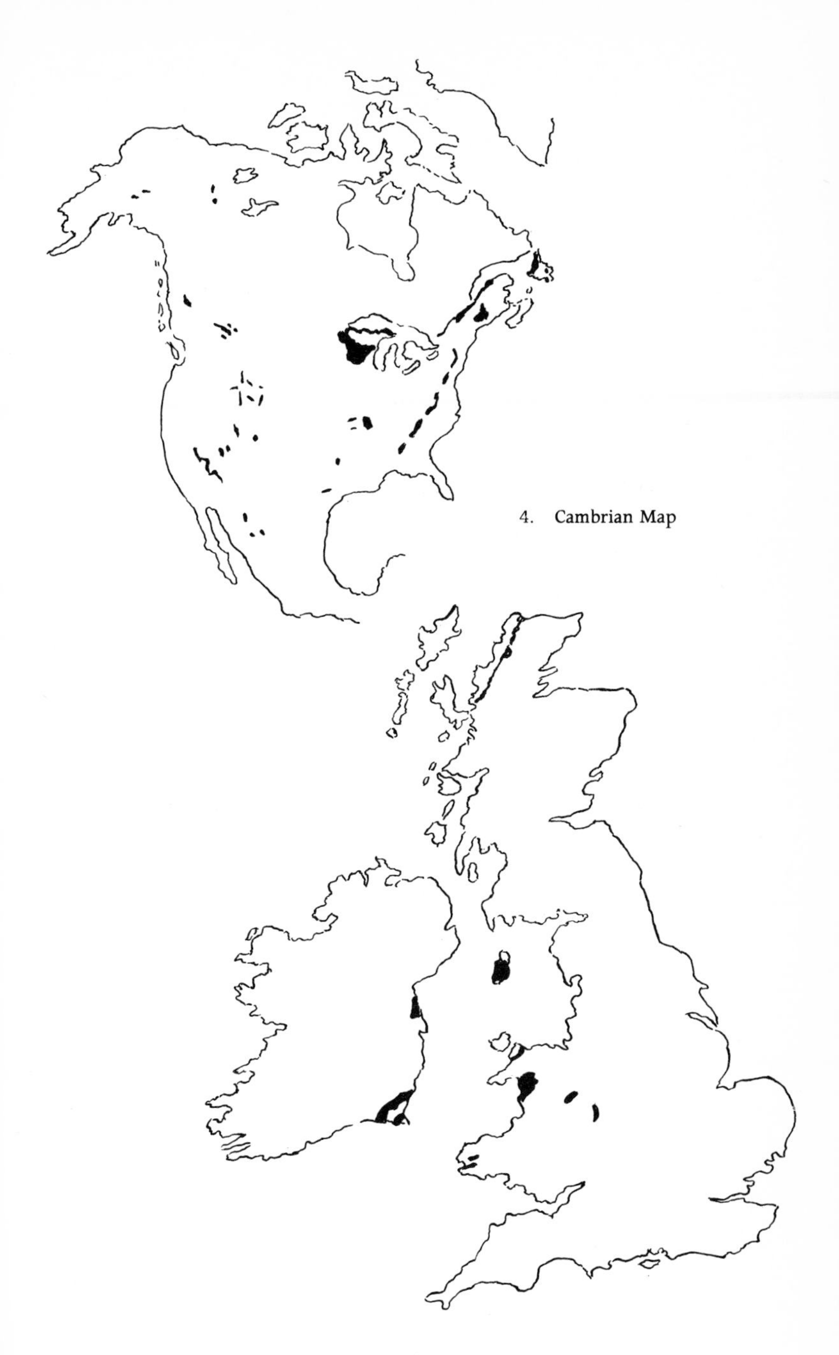

4. Cambrian Map

Cambrian System the Trilobites were the dominant organisms, but there were also Jellyfish, Arachnids, Crustaceans, Brachiopods, Worms and Sponges. Many of these occur in the Burgess Shale of British Columbia, Canada. In the Upper Cambrian System the Pelecypods or Bivalves also occurred and in the Tremadoc Slates, Wales, Graptolites occurred for the first time. These organisms had no internal skeleton, being Invertebrates, but some had a hard shell-covering or skin for protecting the soft parts, and it is this which became fossilized. Also occurring, all in the sea, in the Cambrian Period were Protozoa, Foraminifera, Crinoids, Sea Cucumbers.

THE ORDOVICIAN PERIOD (Fig. 5) rocks–Sandstones, Mudstones, Limestones, Slates and Shales–contain Trilobites, Graptolites, Echinoids, Asteroids, Bryozoans, Crinoids, Brachiopods, Gastropods, Pelecypods and Cephalopods, all sea-living Invertebrates. An exception is the *Astraspis*, an Ostracoderm Fish, whose tiny fragments of hard outer armour and bony backbone were found in Colorado, U.S.A., possibly the first evidence of a Vertebrate, an organism with a backbone. The first Corals also occurred during this Period. Outcrops of Ordovician fossil-

5. Ordovician Map

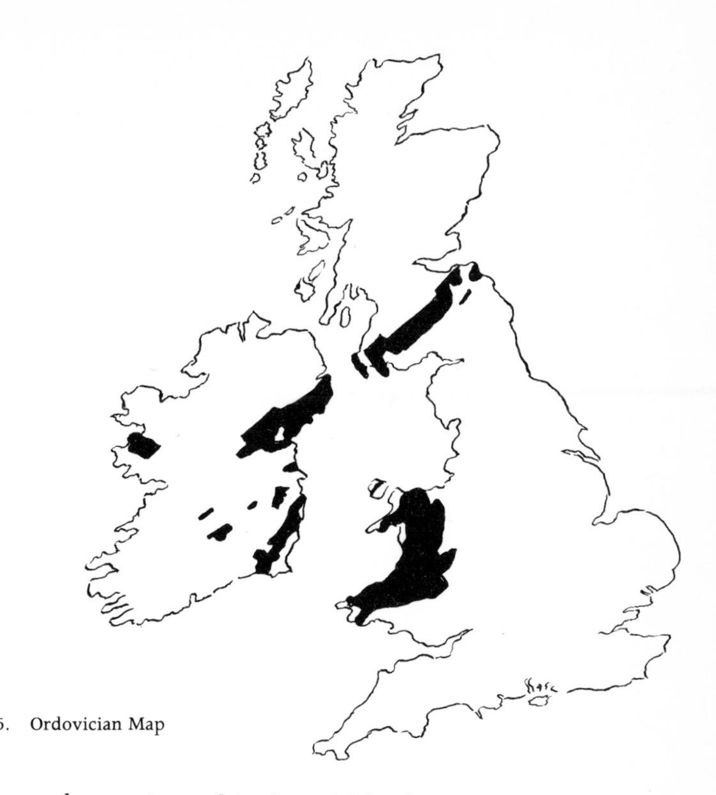

5. Ordovician Map

bearing rocks are situated in the British Isles–in Wales, Scotland, Lake District, Shropshire, and Ireland; in the U.S.A. and Canada in widely scattered areas, including Quebec, Vermont, Wyoming and Ohio.

THE SILURIAN PERIOD (Fig. 6) is notable for two events. One is the appearance of plant life on land, as well as in the sea. Some of the oldest fossils of land plant life were found in Silurian rocks in Australia. Secondly, several primitive, but recognisable, Fish occurred in the Silurian seas and were fossilized. An example is *Birkenia,* an Agnathan Fish, about 4 inches in length, which did not have true jaws, found in Silurian rocks in Scotland. Another sea-inhabitant of this Period was the Invertebrate *Eurypterid,* resembling a cross between a Scorpion and a larval staged Crab, which measured up to 9 feet in length and could swim powerfully. Corals also increased in number during this

6. Silurian Map

Period. There were also Algae, Trilobites, Graptolites, Cephalopods, Brachiopods, Crinoids, Starfish, Gastropods. The Silurian rocks, Limestone, Sandstone, and Shales, occur in the northern and eastern U.S.A. and Canada, the British Isles, Scotland, Wales, the Lake District, Shropshire, and Ireland.

THE DEVONIAN PERIOD (Fig. 7) also saw two important occurrences. One was the development of the first land Vertebrate animals, primitive Amphibians, *Ichthyostegids*, from 'fringe-fins', air-breathing Fishes. The second concerned the oldest known Insects, Spiders, Scorpions, Millipedes. The variety and number of Fishes increased. These included the jawless *Ostracoderms; Placoderms*, armoured Fishes with bony plate-skins; the ancestors of the Sharks; and *Osteichthyes*, the first Fishes with jaws and bony skeletons. From the latter developed the *Crossopterygians*, the lobe- or 'ray-fin' Fishes, and the 'fringe-fin' Fishes, which had fins on bony stalks and a type of air-bladder and nostrils so they could breathe. Land Plants also increased in number and by the late Devonian Period there were Tree Ferns, Seed Ferns, and Horse-tails. Corals, Sponges, Crinoids, Brachiopods, Euryp-

7. Devonian Map

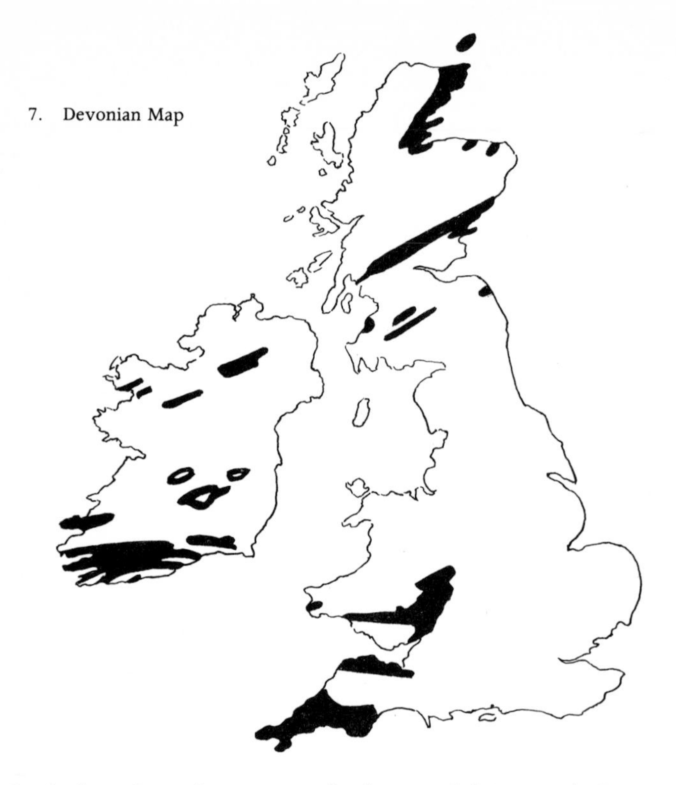
7. Devonian Map

terids, Cephalopods and Gastropods, but Trilobites and Graptolites began to die out in this Period. It was during the Devonian Period that the Old Red Sandstone rock was laid down in areas of Wales, Scotland and Ireland, but it contains only a few fossils.

THE CARBONIFEROUS PERIOD (Fig. 8) (British term–see North America Geological Time Scale) is usually divided into two sections because it contained a variety of occurrences resulting not only in an abundance of organisms, but also changes in sedimentation. The early part of the Carboniferous Period, the Lower, was one of shallow, warm seas containing lime. Mud formed which became white Limestone rock. This contains large numbers of fossil Crinoids, Corals, Brachiopods, the first coiled Cephalopods and Gastropods. Following this there were changes in sedimentation; Millstone Grit, Shales and Sandstone were laid down in some areas of the British Isles, in which there are few fossils, usually Plants and Goniatites, an early form of Ammonite. During the second half of this Period, the Upper, a change occurred in which the sea-bed was raised to create low-lying land and on it lagoons and swamps came into existence. This allowed the vast Coal Forests, with Trees (*Lycopods*), seed Ferns (*Pteridosperms*), Horse-tails and Clubmosses to grow. As they died and rotted these were gradually changed into Coal deposits and from this the Period got its name. The swamps also allowed the increase of Amphibians, which preyed upon the air-breathing Fishes and King Crabs, or the Gastropods, Scorpions and Molluscs that inhabited the Ferns and Clubmosses. During this Period some of the Insects were winged and were able to fly. One, *Meganeura*, resembled a large Dragonfly, with a 30-inch wing span, and was an inhabitant of the swamps. A very important development was the change made by the Amphibians, which, instead of laying their eggs in pools of water for their early life stages, laid them, enclosed in a shell for protection, on raised land. From these developed the first Reptiles.

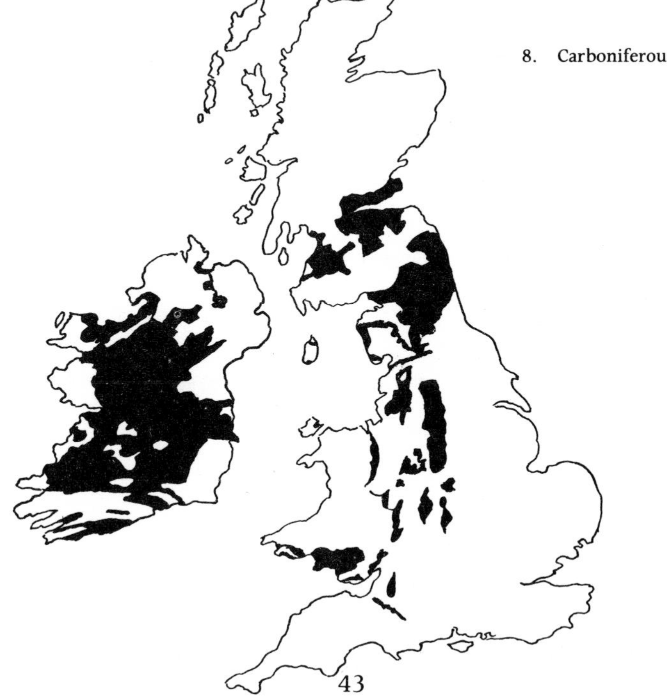

8. Carboniferous Map

9. Mississippian Map

THE MISSISSIPPIAN PERIOD (Fig. 9) (North American term–see North America Geological Time Scale) corresponds with the Lower Carboniferous Period in the British Geological Time Scale. It was so-named because of the Limestone outcrops of the Mississippi River. In the Period, Corals, Crinoids, Brachiopods, Bryozoans and Foraminifera occurred abundantly, with Amphibians and Land Plants, prior to the occurrence of the Coal Forests.

THE PENNSYLVANIAN PERIOD (Fig. 10) (North American term–see North America Geological Time Scale) corresponds with the Upper Carboniferous Period in the British Geological Time Scale. It was so-named after the coal seams in Pennsylvania. In this Period, Corals, Crinoids, Brachiopods, Bryozoans,

Foraminifera and Gastropods occurred abundantly, while in the low-lying swamps, Trees, Ferns, Horse-trails and Clubmosses, as in the British Isles, died and consolidated into coal; giant Insects occurred, also numerous Fishes and from the many Amphibians developed the Reptiles.

10. Pennsylvanian Map

THE PERMIAN PERIOD (Fig. 11) could be called the beginning of the Reptile Age. An intermediate stage is the *Seymouria,* half-Amphibian, half-Reptile, which had a large, flat head and measured about 2 feet in length. True Reptiles were the scaly, carnivorous sail-back Lizards, such as *Dimetrodon,* which had tall fin-like extensions along their backs and measured up to 10 feet in length. *Mesosaurus* was an 18-inch, long-snouted aquatic Reptile with needle-like teeth. The *Theriodonts* were small carnivorous Reptiles from which Mammals descended. Although new Insects, including Beetles and true Dragonflies appeared, many of the Animals, Plants and organisms which had prominently existed in earlier Periods now declined or became extinct. Amphibians and Fishes began to decline in face of the sometimes viciously savage Reptiles and some of the Bryozoans, Corals, Brachiopods, Foraminifera, Crinoids and Cephalopods disappeared, while many of the Ferns and Horse-tails also died out and the Coal Forest Trees and Plants declined, to be replaced by primitive Conifer Trees. The Magnesian Limestone is the best sediment rock for fossils of this arid Period, and occurs in Durham, Yorkshire, and the Midlands in the British Isles, while Permian rock outcrops occur in western and northern U.S.A., especially Texas, and in Canada.

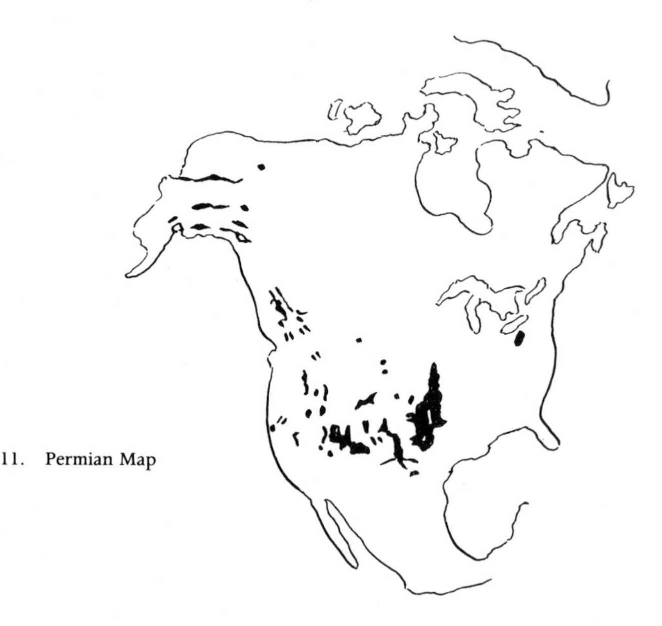

11. Permian Map

THE MESOZOIC ERA

THE TRIASSIC PERIOD (Fig. 12) is thus named because of the distinct three-fold layers of its rocks. During this arid Period, scaled Reptiles increased their domination on land and in the sea. The *Ichthyosaurus*, meaning Fish-Lizard, was a sea Reptile with a streamlined, Dolphin-like body, a fin on its back, a beak-like head armed with teeth, paddle-like feet and a tail like the Fish it preyed on. The carnivorous *Triassochelys* was the first

11. Permian Map

primitive Turtle, which appeared in this Period. On land the *Cynognathus* was a 7-feet long carnivorous Mammal-like Reptile which had a Dog-like skull and sharp Dog-teeth that it used to tear a victim apart. Those organisms which had disappeared were replaced by new types of Brachiopods, Corals, Sponges, and Crinoids. Ammonites were common. Gastropods and Pelecypods increased again, although the Trilobites were now extinct. Amphibians and air-breathing Fish were becoming rare, but boned Fishes increased numerically and a new development was the Flying Fish. Some of the present-day Echinoids and Crinoids now appeared. The Primitive Conifers also thrived, while various Ferns survived in the remaining wet land areas. Red Sandstones and Marls were the main Triassic rocks deposited and these do not contain many fossil remains.

12. Triassic Map

THE JURASSIC PERIOD (Fig. 13) was one in which the famous *Dinosaurs*, which means 'terrible lizards', roamed the Earth. Some Reptiles were huge monsters, one of these being the herbivorous *Diplodocus*, with a long, slender neck and tail, which altogether measured 87 feet in length. Others were the herbivorous *Brontosaurus*, up to 67 feet long; the armoured herbivorous *Stegosaurus*, with a frill of bony plates along its back, a spiked tail, and a tiny-brained small head, altogether measuring 20 feet and weighing 10 tons; the carnivorous *Allosaurus*, 35 feet in length, which stood and moved on its hind legs; the marine *Plesiosaurus*, with a long, slender neck and paddle-like legs, up to 40 feet long; and the *Eurhinosaurus*, with a long, armed snout, like a Swordfish in appearance, up to 17 feet long. There were also winged Reptiles, such as *Rhamphorhynchus*, with

Plate III *Belonostomus munsteri* Agassiz. Lithographic Limestone, Jurassic. Solenhofen, Bavaria.

Plate IV Swimming Crab. Tertiary.
Patagonia.

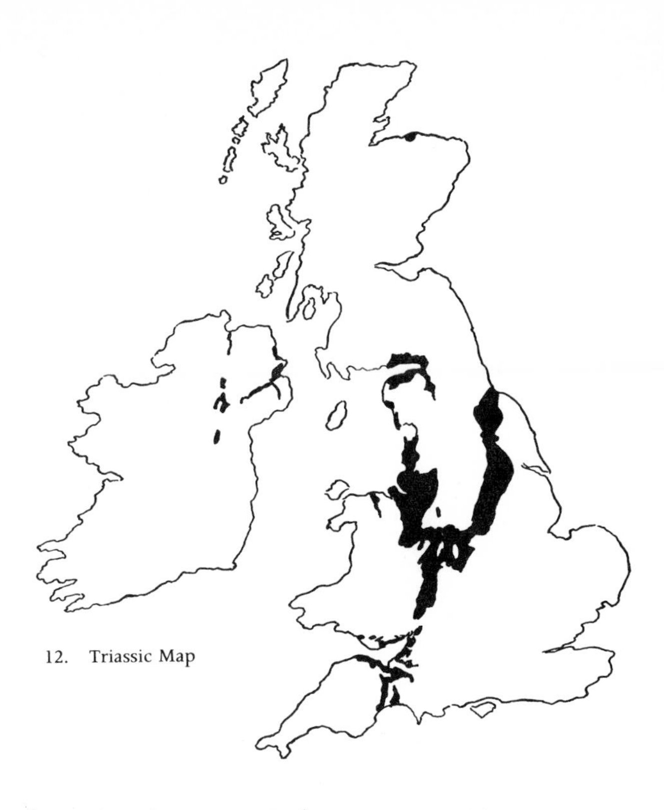

12. Triassic Map

slender, pointed wings and a long tail. The oldest known Bird, the *Archaeopteryx*, also occurred in this Period. There was also a varied range of other organisms-Pelecypods, Gastropods, Brachiopods, Echinoids, Corals, Crinoids, Foraminifera, Belemnites, Fish, Sharks, Rays, Turtles, while coiled Ammonites were abundant, varying in size from 1 inch to 6 feet in diameter. There was also a population of Insects–Beetles, Flies, Grasshoppers and Dragonflies. Plant life of the Period included Conifers, Ferns, Ginkgos and Palm-like, cone-bearing Cycads. The Jurassic rock deposits consist of Limestone, Sandstone, Shale, Slate, Marl and Clay. The deposition is divided into three sequences–the first or Lower, the Middle and the last or Upper, which can be dated by the fossils they contain. Marls were generally first deposited, followed in the late Jurassic Period by Sandstones and Limestones.

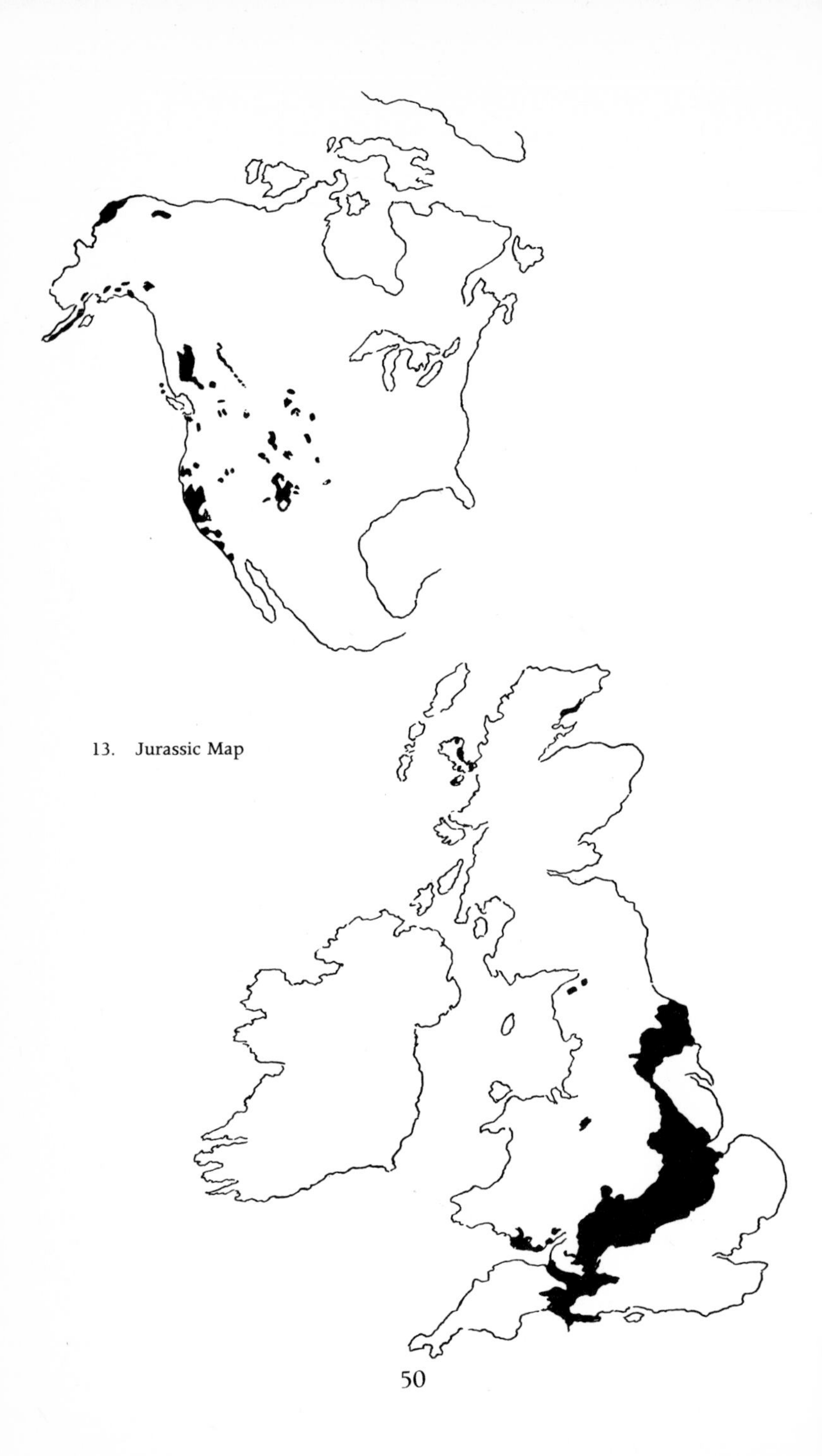

13. Jurassic Map

THE CRETACEOUS PERIOD (Fig. 14) saw the Age of Reptiles come to its conclusion and by its end they had almost vanished. Even so, from the early part of the Period until this eventuality there were still gigantic monster-like Reptiles on land and in the sea. One was the carnivorous *Tyrannosaurus*, 'tyrant lizard', which stood 20 feet high on its massive hind legs holding its very small forelegs like arms, and measured 50 feet in length. Similar was the herbivorous *Iguanodon* which balanced its ungainly body by means of its long, thick tail. A fearsome marine Reptile was the carnivorous *Tylosaurus*, with a Lizard-like body and toothed pointed jaws, up to 26 feet long. The *Archelon* was a marine Turtle, 12 feet in length. In the skies flew the *Pterodactyls*, 'wing fingers'; one of them, the toothless *Pteranodon*, had a wing span of 25 feet. There were also primitive Birds, one being the *Ichthyornis*, which stood about 8 inches high and had a similarity to the modern Tern. In the Cretaceous seas there was a great variety of organisms, which became fossilized– Echinoids, Crinoids, Gastropods, Pelecypods, Sponges, Corals, Foraminifera, Belemnites. Ammonites continued in abundance. The most significant event of the Period, however, was the appearance of the first Flowering Plants (*Angiosperms*) on the Lower Cretaceous land. With several other factors, such as geological changes and climatic changes this meant the downfall of the Reptiles. As the Flowering Plants became dominant and Trees, such as Oak, Elm, Maple, Holly, Beech, Poplar, appeared, the evergreen Cycads and the herbivorous Reptiles which fed upon them disappeared. This in turn meant the decline of the carnivorous Reptiles. By the end of the Cretaceous Period the *Pteriosaurs, Plesiosaurs, Pterodactyls, Ichthyosaurs, Dinosaurs, Mesasaurs*, the *Ammonites, Belemnites*, were extinct.

A familiar rock of the Period is the unique, pure, fine-grained Limestone known as Chalk, which contains large numbers of fossils. The deposition of the Chalk is divided into three sequences, the first or Lower, the Middle, and the last or Upper. These can be dated by the fossil types contained within them. Early in the Period there was a deposition of mud which formed the muddy Blue Gault and Greensand, the latter named because it contains the green mineral glauconite. Both of these also have in them considerable quantities of fossils, particularly Ammonites. In the Chalk there are layers of Flint, made up of spicules of silica, and in these stones fossils, particularly Echinoids, are sometimes found enclosed.

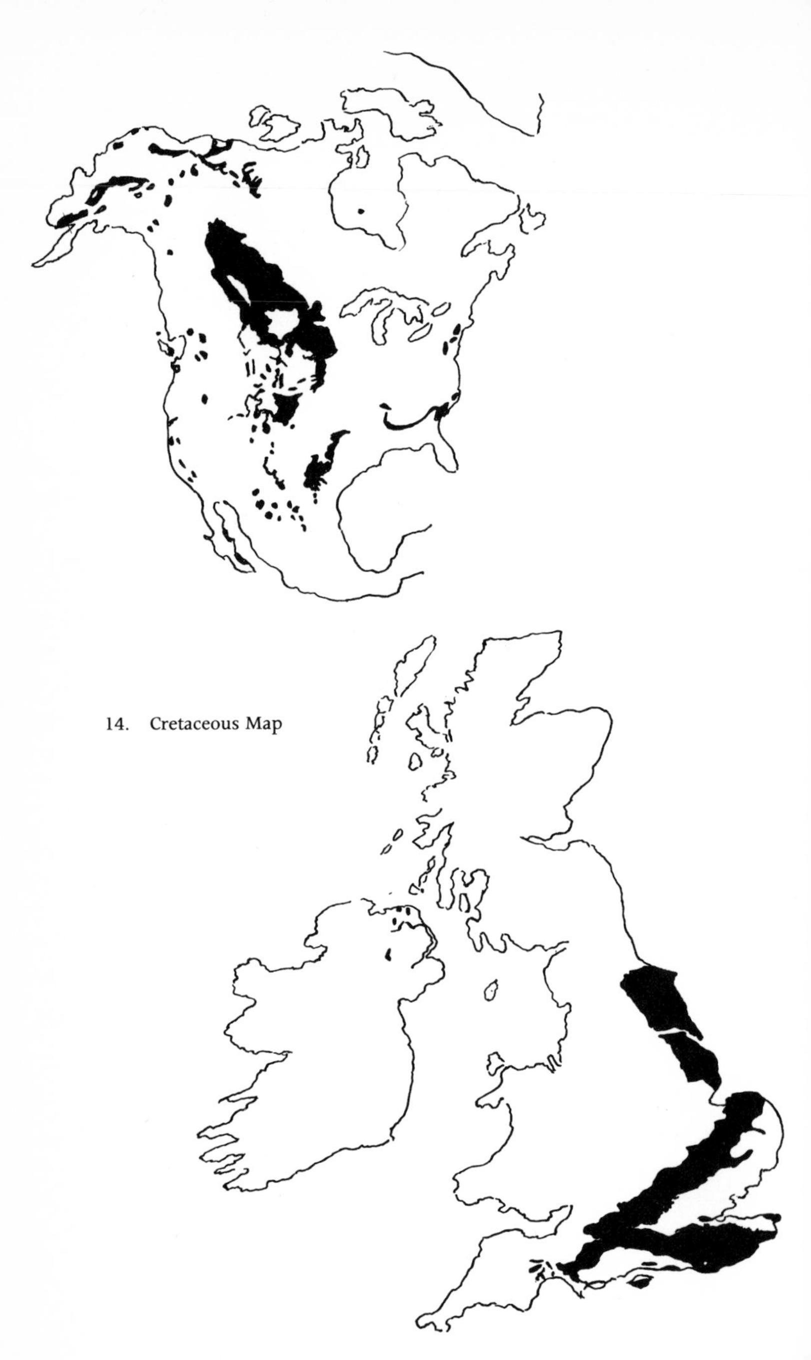

14. Cretaceous Map

THE CAINOZOIC [CENOZOIC] OR TERTIARY ERA

THE PALAEOCENE PERIOD, THE EOCENE PERIOD AND THE OLIGOCENE (Fig. 15) comprise the early or Lower part of the Tertiary Era. During these Periods the Age of Mammals was founded with the entire decline of the prehistoric monster-type Reptiles. Many of the Mammals and Flowering Plants of these Periods and the Upper Tertiary which appeared, have survived with virtually little change until the present time. The ancestors of some of the modern Mammals and Birds, however, were different.

In the Eocene Period the *Moeritherium*, the ancestor of the Elephant had a pointed snout, no trunk, and stood about the height of a Pig. The herbivorous *Eohippus*, an ancestor of the Horse, had 4 toes and stood 12 inches high, no more than a Terrier. There were also huge Mammals and Birds which appeared, then became extinct. In the Eocene Period the herbivorous *Uintatherium* was a large-hoofed Mammal, about the size of a Rhinoceros, which had three pairs of blunted horns projecting from its nose to its ears. The carnivorous *Diatryma* was a 7-feet-tall bird. The ancestors of the true Rodents and Bats appeared, so did the *Condylarths*, ancestors of hoofed Mammals, and the *Creodonts*, primitive ancestors of our carnivorous Mammals; but the latter two, *Condylarths* and *Creodonts*, eventually disappeared and were replaced by more successful Mammals. The first of the Cats, *Hoplophoneus*, a sabre-tooth the size of a Cougar appeared. Whales were present in the sea and modern Fish abounded. Plant and Tree life included Palms, Swamp Cypresses, Pines, Magnolia, Plane and Hazel; and Water Lilies flourished in the lakes. Marine life included numerous modern Molluscs, Oysters, Cowries, also Crabs and Starfish. The once-abundant Ammonites were now almost extinct.

In the Oligocene Period, the *Mesohippus*, another ancestor of the Horse, with three toes and standing about 24 inches high, appeared. So did the herbivorous *Baluchitherium*, an 18-feet-high Rhinoceros, but it died out. The ancestors of the Tapirs, *Protapiris*, now occurred, also a land Tortoise, the *Stylemys*. In this Period the first Grasses appeared and became widespread.

In the British Isles the Palaeocene rocks, such as the Thanet Sands, contain many fossils, especially Shells. The Eocene Period rocks also contain numerous fossil Shells, the fossiliferous

London Clay, a thick deposit of clay, being of this Period. The Oligocene Period rocks in the British Isles, especially the Isle of Wight, contain marine life and Mammal fossils. Fossils of these Periods occur in the south-eastern and western regions of North America.

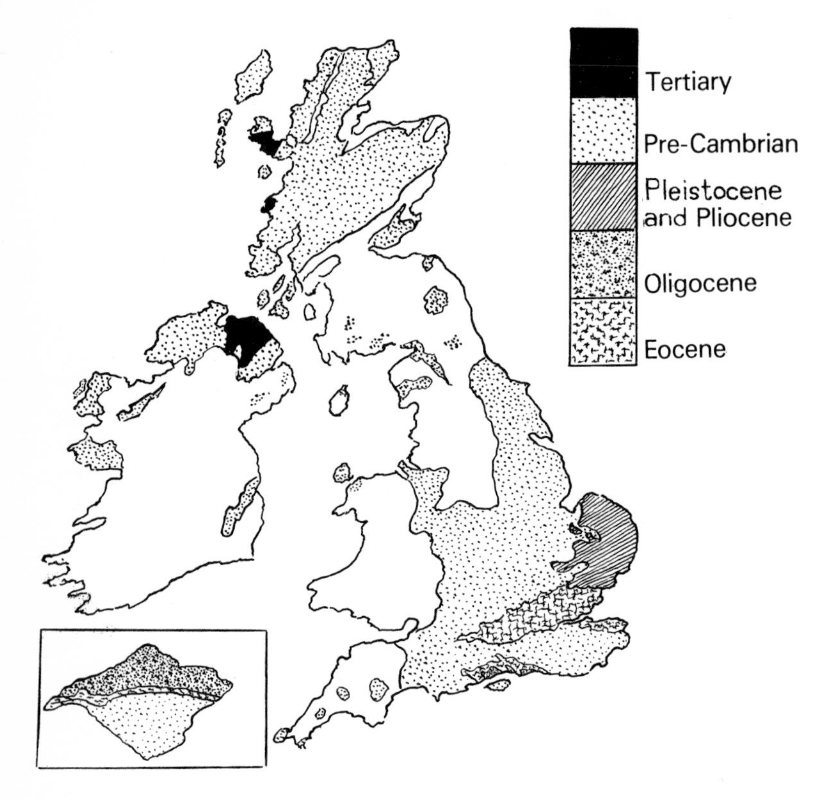

THE MIOCENE PERIOD AND THE PLIOCENE PERIOD comprise the later or Upper part of the Tertiary Era. During these Periods there was a continual development of Mammals, some of these surviving, others disappearing, depending on changes in climate which in turn affected vegetation. When these caused the disappearance of Forest to be replaced in the Miocene Period by Grassland, so the Mammals which browsed were affected and replaced by grazing Mammals. In the Miocene Period the

Merychippus, a larger 3-toed ancestor of the Horse, 40 inches in height, appeared, to be followed in the Pliocene Period by the *Pliohippus*, a 1-toed ancestor of the Horse 50 inches in height and resembling the Horse we know today. In the Miocene Period and Pliocene Period there were also *Dinohyus*, a giant Pig; *Stenomylus*, an ancestor of the Camel; *Diceratherium*, an ancestor of the Rhinoceros. The herbivorous *Oreodon*, about the same size as a Sheep, died out in the Pliocene Period. *Moropus*, like a huge long-necked Horse with claws also appeared and disappeared. There were also Dogs and various small carnivores; an ancestor of the Elephant, the *Trilophodon*, a Mastodon with 4 tusks; Giraffes, Antelopes; but the Amphibians and Reptiles were generally reduced in size as we know them, while marine life was in forms still present today, and so was the Plant and Tree vegetation. Most significant to us is the small Ape living in Forest Trees in Africa in the Miocene Period. Known as *Proconsul*, from it descended all the modern Apes–and Man.

There are no Miocene Period rocks in Britain, but Pliocene Period deposits do occur as 'Crags', or Shelly Sands, in East Anglia. Here may be found various fossil Shells, Sharks' teeth, and vegetation. Rocks containing fossils of these Periods occur in North America.

15. Tertiary Map

THE QUATERNARY ERA

THE PLEISTOCENE PERIOD contained four glacial and inter-glacial sequences, where vast ice sheets covered much of Europe and North America. During these sequences there was considerable migration of wild life. The 12-foot-tall Woolly Mammoth and 6-foot-tall Woolly Rhinoceros moved to find the tundra, Mosses, Lichen, Dwarf Birch and Arctic Willow; while carnivores, such as the *Smilodon,* a sabre-toothed Tiger, Wolves and Arctic Foxes, followed them as prey. Pigs, Bison, Reindeer, Camels, Elephants, Badgers, Elk, and Deer, etc., roamed according to the climatic changes. Some of these died and were preserved in the tar lakes. Fossilized remains of Coniferous and Deciduous Trees and Plants of this Period have been found in the British Isles and North America, also Fish and Mammoth Teeth, Deer Antlers, and Gastropod Shells.

THE HOLOCENE OR RECENT PERIOD is the time we live in. Its commencement saw the retreat of the ice sheet and the extension of modern-type Forest. To fossil collectors it is not of particular importance, except that preserved Tree and Plant spores, pollen, leaves, stems and branches have been found.

The Geological Time Scales

The Geological Time-Scale and Stratigraphical Column of the British Isles

PRE-CAMBRIAN OR PRE-PHANEROZOIC ERA

	Series	*Approximate Duration in Million Year Units*	*Period Commencement from Present Time in Million Year Units*
Uriconian (England):	Malvernian Longmyndian	4,030	
Mona Complex (Wales)	The Padarn Ridge		4,600
Dalradian (Scotland)	Lewisian Torridonian Moinian		

LOWER PALAEOZOIC ERA

Period or System	*Series*	*Approximate Duration in Million Year Units*	*Period Commencement from Present Time in Million Year Units*
Cambrian	Lower Cambrian Middle Cambrian Upper Cambrian Tremadoc	70	570
Ordovician	Arenig Llanvirn Llandeilo Caradoc Ashgill	70	500
Silurian	Llandovery Wenlock Ludlow Downtonian	35	430

UPPER PALAEOZOIC ERA

Period or System	*Series*	*Approximate Duration in Million Year Units*	*Period Commencement from Present Time in Million Year Units*
Devonian	Lower Devonian or Lower Old Red Sandstone Middle Devonian or Middle Old Red Sandstone Upper Devonian or Upper Old Red Sandstone	50	395
Carboniferous	Carboniferous Limestone Millstone Grit Coal Measures	65	345
Permian	Magnesian Limestone Permian Marls	45	280

MESOZOIC ERA

Period or System	*Series*	*Approximate Duration in Million Year Units*	*Period Commencement from Present Time in Million Year Units*
Triassic	Bunter Beds Keuper Beds Rhaetic Beds	55	225
Jurassic	Lower Lias Middle Lias Upper Lias Inferior Oolite Great Oolite Cornbrash Oxford Clay Ampthill Clay Corallian Beds Kimmeridge Clay Portland Beds Purbeck Beds	35	190
Cretaceous	Speeton Clay Red Chalk	70	135

	Series		
	Wealden Beds Lower Greensand Upper Greensand and Gault Lower Chalk Middle Chalk Upper Chalk		

CAINOZOIC OR TERTIARY ERA

Period or System	*Series*	*Approximate Duration in Million Year Units*	*Period Commencement from Present Time in Million Year Units*
Palaeocene		11	65
Eocene	Thanet Sands Woolwich Beds Reading Beds London Clay Bagshot Beds Bracklesham Beds Barton Beds	16	54
Oligocene	Headon Beds Osborne Beds Bembridge Beds Hamstead Beds Bovey Tracey lignites and clays	12	38
Miocene		19	26
Pliocene	Lenham Beds Coralline Crag	4·5	7

QUATERNARY ERA

Period or System	*Series*	*Approximate Duration in Million Year Units*	*Period Commencement from Present Time in Million Year Units*
Pleistocene	Red Crag Norwich Crag Cromer Forest Bed		2–5
Holocene or Recent			5,000 B.C.

The Geological Time-Scale of the U.S.A. and Canada

It will be noted that the North American continent Geological Time-Scale varies in some respects to that of the British and European Time-Scale, namely in the Period commencement and duration. In some respects geologists are not entirely in agreement with the Time-Scales and Stratigraphical Columns in use concerning Period definition and so these disparencies have occurred and I have given those which are relevant either to the reader in North America or the British Isles. The uranium method, based on the radioactive element breakdown, also others using rubidium, thorium, potassium and carbon, can measure the age of a rock and the fossils it contains accurately in years and so the production of a Geological Time-Scale indisputably accurate is now possible and it is to be hoped will shortly be realised. In the North American Time-Scale the Mississippian and Pennsylvanian Periods in the Palaeozoic Era correspond similarly with the Carboniferous Period, in the Upper Palaeozoic Era, in the British Time-Scale.

PRE-CAMBRIAN ERA

	Period Commencement from Present Time in Million Year Units:
	4,500

PALAEOZOIC ERA

Period or System	*Approximate Duration in Million Year Units*	*Period Commencement and length from Present Time in Million Year Units*
Cambrian	100	600–500
Ordovician	75	500–425
Silurian	20	425–405
Devonian	60	405–345
Mississippian	35	345–310
Pennsylvanian	30	310–280
Permian	50	280–230

MESOZOIC ERA

Period or System	*Approximate Duration in Million Year Units*	*Period Commencement and Length from Present Time in Million Year Units*
Triassic	49	230–181
Jurassic	46	181–135
Cretaceous	72	135–63

CENOZOIC (CAINOZOIC, TERTIARY) ERA

Period or System	*Approximate Duration in Million Year Units*	*Period Commencement and Length from Present Time in Million Year Units*
Palaeocene	5	63–1
Eocene	22	
Oligocene	11	
Miocene	12	
Pliocene	12	

CENOZOIC (QUATERNARY) ERA

Period or System	*Approximate Duration in Million Year Units*	*Period Commencement and Length from Present Time in Million Year Units*
Pleistocene	1	
Recent		

Quick reference Glossary of some text terms

Ambulacral Applies to the plates of the *Echinoids, Crinoids* and *Blastoids,* the rows of ambulacral plates being those on the outside of the Sea-Urchin test or upper edge of the *Blastoids* and *Crinoids* cup.

Beak The curved, protuberant apex tip of a Bivalve Shell. Also called the *Umbo.*

Bed A large mass of material, Limestone Bed, Sandstone Bed, etc., sometimes containing fossils.

Calcareous Containing lime or limestone.

Carapace The thick upper shell of a Crab or other Crustacean, a Turtle or Tortoise.

Cartilaginous Composed of cartilage or resembling cartilage in texture.

Chitin The hard, horny material externally covering *Invertebrates* such as Lobsters, Beetles, etc.

Crag A Shelly Sand.

Crenulated Notched, indented, scalloped edge or edges.

Creodont An archaic Carnivore.

Cusp A projecting point.

Dissepiment A short connecting bar; a dividing partition or wall.

Gneiss Composite rock.

Head Shield Protective, bony covering around the head and fore-part of certain early primitive Fish.

Marl Muddy rock.

Marsupial A Mammal with a pouch for carrying young–Kangaroo, Wombat, etc.

Node A protuberance or joint on a plant stem.

Pelagic Inhabits the open deep sea.

Periostome Circular area surrounding the mouth of Sea Urchins.

Pinnules Divisions of leaflets of a pinnate leaf.

Placental Relating to the *placenta,* the part of a seed vessel to which the ovules are attached or tissue mass attached to a womb.

Plicated Having folds or ridges.

Schist A crystalline, metamorphic rock formation, which splits easily into plates.

Siphon Canal A groove in the lip or extension of the aperture's margin of a *Gastropod* shell, where the siphon protruded.

Spicules Small, hard granules found in Invertebrates, such as forming the skeleton of a Sponge.

Stipe A stalk, stem, or branch.

Striae A channel, groove; a thread-like or flute-like marking.

Sulcus A prominent depression or long narrow groove.

Suture The continuous spiral line of a *Gastropod* shell which is created by the adjoining junctions of the whorls.

Thecae The cups attached to the *stipes* or stems of *Graptolites.*

Umbo The area of a Bivalve shell sited behind the *Beaks,* but is also used in descriptions to include the *Beaks,* or as an alternative to *Beak.*

Venter The abdominal cavity of Insects and other *Invertebrates.*

Ventral The underside or outer side.

FOSSIL EXAMPLES
Part One
Vertebrates

AVES (Birds) [Vertebrates]

The first birds developed from the two-footed Reptiles. Although these still had reptilian characteristics, they were true birds with feathers replacing the Reptiles' scales. Many of the early birds were fish eaters and thus lived in water or close to it. Consequently a large proportion of fossil bird remains are of known water birds. As examples *Argillornis* and *Odontopteryx,* occurring in the Eocene, were similar to present-day Cormorants. *Proherodius,* from the Eocene Period, was a Heron, *Diomedea* was an Albatross-like bird of the Pliocene, *Aetornis,* a Cormorant, and *Ibidipsis,* an Ibis, both from the Oligocene Period. Other known fossil birds are *Lithornis,* from the Eocene-London Clay, and *Palaeocircus,* from the Oligocene–both birds of prey; *Gastornis,* a long-legged land bird from the Eocene; *Phororhacos,* a large, flightless land bird from the Miocene; *Enalliornis,* unknown bird type from the Cretaceous-Greensand; and *Halcyornis,* possibly a Kingfisher-type bird, from the Eocene Period.

Due to their frail, air-filled bones and mode of existence, a lesser number of Birds have so far been discovered as fossils compared with other Vertebrates.

ARCHAEOPTERYX (Fig. 16)

Jurassic. Examples have occurred in the Solenhofen Limestone, Bavaria, Germany. The oldest known bird; its skeleton shows its Reptile ancestry, by the shape of the skull, its teeth, and long, bony, lizard-like tail which was an extension of the vertebral column. Each wing had three finger-like claws and long flight feathers, the body and tail bones also having long feathers. The clawed feet and legs were able to perch. It was about the size of a Crow, 18 inches in length.

16. Skeleton of *Archaeopteryx* bird

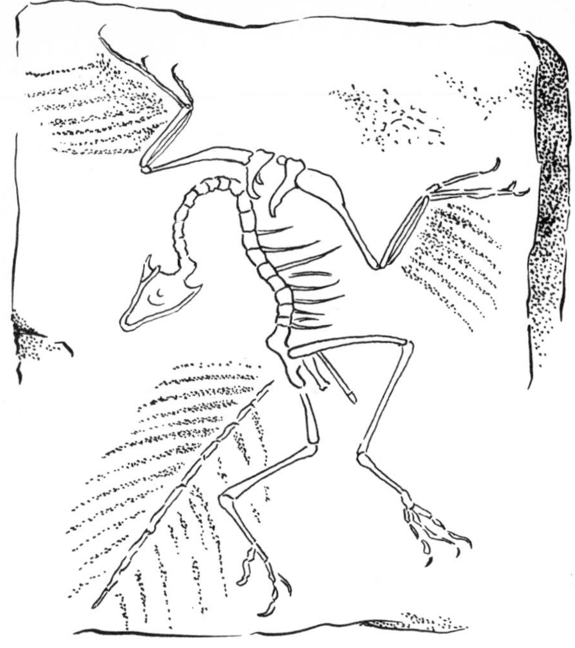

HESPERORNIS

Cretaceous-Kansas, U.S.A. A toothed, flightless bird, with only tiny rudimentary wings; it was a sea inhabitant and used its powerful legs and feet to dive, swim and pursue fish. It had a similarity to the modern Cormorant, but was about 6 feet in length.

MAMMALIA (Mammals) [Vertebrates]

Although the first true Mammals, Shrew-like in appearance, are known from the Jurassic Period and the first Marsupial Mammals and Placental Mammals appeared in the Cretaceous Period, it was not until the dominance of the Reptiles had come to an end, by the close of the Mesozoic Era, that Mammals had the freedom in the Eocene Period to evolve and spread without restriction. In addition to the small Rodent-like Mammals, other forms increased in height to Dog-size and were either herbivorous or carnivorous as opposed to earlier forms being insectivorous. They had small brains, short legs with five-toed feet, plantigrade or flat-footed, and are referred to as Archaic Mammals because of their low stature. These mostly disappeared, however, and the ancestors of specialised Modern Mammals appeared. Some of these in the Eocene and Oligocene Periods were also small but others were huge.

It is not the intention of my book for reasons of space, to describe the ancestral evolution of Mammals, Amphibians and Reptiles through the Periods as there are other specialised books on the subject, but the following are a few examples of Mammal fossils which have been discovered. *Didelphys,* a marsupial of the Eocene, ancestor of the Opossum; *Coryphodon,* a 3-feet-high herbivore in the Eocene Period; *Lophiodon* and *Palaeotherium,* ancestors of the Tapirs, from the Eocene and Oligocene Periods; *Glyptodon,* a 9-feet-long, bony armoured ancestor of the Armadillos, from the Pleistocene Period; *Oxyaena,* a carnivorous creodont, 3 feet in length, in the Palaeocene and Eocene Periods; *Megatherium,* a 20-foot-long Ground Sloth of the Pleistocene Period.

In the following Miocene and Pliocene Periods, the web-footed carnivores, Pinnipeds, such as the Seals and Walruses, evolved, probably from dog-like ancestors. There were also near-ancestors of present-day animals, such as the *Hipparion,* about the size of an Ass, a 3-toed ancestor of the Horse in the Pliocene Period, remains of which have been found below the Red Crag, and the *Mastodon,* a smaller version of the Elephant, with grinding teeth, less ridged, from the Miocene Period.

The fossil remains of Mammals in Britain are mainly of teeth, jaw, skull and other bones, and antlers, but in North America complete Mammal fossil remains, such as the *Eohippus,* another ancestor of the Horse, have been discovered. Should there be found in Britain an entire or almost entire fossil skeleton of a Mammal, because of its scarcity, expert help should immediately be sought, not only for the best method to recover it from its enclosing material, but also for its precise identification.

HIPPOPOTAMUS AMPHIBIUS (Fig. 17)
Pleistocene-Cromer Forest Bed, British Isles; Europe. An ancestor of the present-day Hippopotamus, which lived a similar existence in rivers and freshwater lakes. Fossil remains

in British Isles, teeth, chiefly molars, these having irregular edges and flattish biting surface.

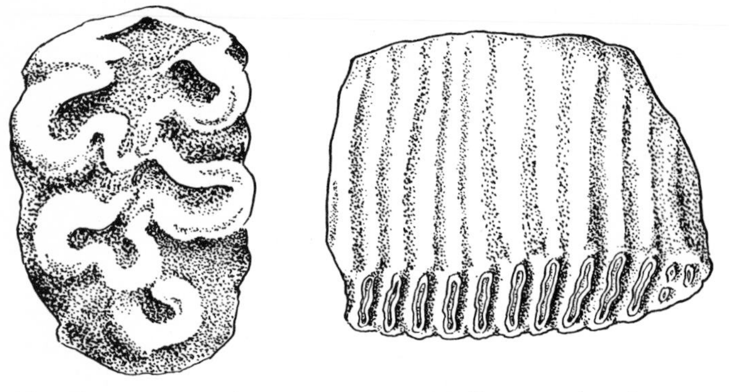

17. Hippopotamus tooth   18. Woolly Mammoth tooth

IRISH GIANT DEER [ELK] (Megaceros [Cervus] giganteus)
Pleistocene-Ireland. The largest known Deer; its antlers had a stretch of 11 feet from tip to tip. Skeletons have been found in peat bogs; as also have the antlers.

RED DEER (Cervus elaphus)
Pleistocene-Cromer Forest Bed to Recent. Fossil remains, antlers, sometimes complete or as separated branches. Shape and size unchanged to present-day examples.

REINDEER (Cervus [Rangifer] tarandus)
Pleistocene. Fossil remains, antlers, sometimes complete or as separated branches. Shape and size unchanged from present-day examples.

WOOLLY MAMMOTH Mammuthus [Elephas] primigenius) (Fig. 18)
Pleistocene, British Isles, Europe, North America. A large Elephant, 12 feet in height, that was clothed in long, hairy fur and had very large, upcurving tusks. Body has been found preserved in Siberian frozen mud, but elsewhere occurs as bones and teeth, chiefly molars. Above view, upper molar, egg-shaped, biting surface cross-ridged; sides of consolidated plates forming teeth, ridged.

REPTILIA (Reptiles) [Vertebrates]

The *Reptilia* evolved from a group of the Amphibians, the *Labyrinthodontia.* One of the earliest known examples is *Seymoüria*, a primitive Reptile, with reptilian vertebrae and limbs, but also having Amphibian characteristics, which has been discovered in beds, in Texas, U.S.A., of the Permian Period. It was about 2 feet in length.

The early carnivorous and herbivorous Reptiles, of the Permian and Triassic Periods, were small, but were followed in the Jurassic and Cretaceous Periods by Reptiles not only huge in size, but having structural variations and a grotesque appearance. By the end of the Mesozoic Era the majority of the specialised Reptiles were extinct for reasons given in the Cretaceous Period Section of Chapter Two. Of the then fifteen major Reptile groups only four–the Crocodiles, Turtles, Lizards, Snakes–survive today.

In my book, again for space reasons, it is not intended to describe the evolution of the different Reptile groups through the millions of years in the Periods when they existed, as there are several books available which treat the subject in precise detail, particularly with regard to anatomical differences, but the following are a few of the Reptile fossils which have been discovered.

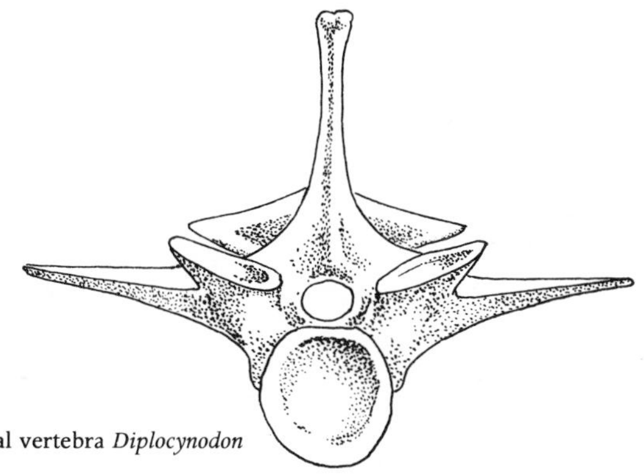

19. Dorsal vertebra *Diplocynodon*

Diplocynodon (Fig. 19), a Crocodile of the Eocene and Oligocene Periods; *Ichthyosaurus,* a carnivorous marine Reptile, up to 30 feet in length, Porpoise-like, with a long toothed jaw, of the Triassic, Jurassic and Cretaceous Periods; *Ornithocheirus* (Fig. 20), a Pterodactyl, winged Reptile, of the Cretaceous Period-Greensand; *Trionyx,* a Turtle, of the Oligocene Period; *Steneosaurus,* a Crocodile of the Jurassic; *Edaphosaurus,* a herbivorous Pelycosaur up to 11 feet long, with a vertebral 'sail' on its back, of the Permian; *Placodus,* a 5-feet-long Mollusc-eating, marine Reptile with a Walrus-like appearance, of the Triassic Period.

The fossil remains of Reptiles in Britain chiefly consist of separate teeth, jaws, vertebrae, limb and other bones, although examples of small Reptiles have been uncovered. Fossil remains of Turtles occur as bones and pieces of the crushed carapace or 'shell'. North America, having a larger land-mass, provides a greater possibility of complete Reptile fossil remains being discovered. If an entire or almost entire fossil skeleton is found, as in the case of Mammal fossils, expert assistance should be sought, not only because such a find could be of prime importance to knowledge of fossils in the locality, but also to obtain precise identification and help to remove the remains in the best way possible for their preservation.

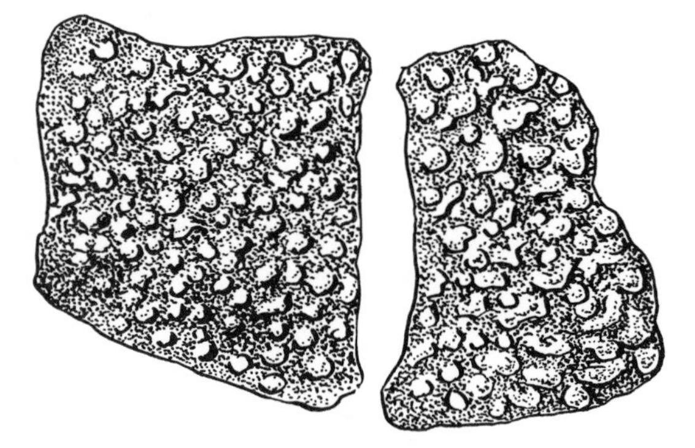

20. Carapace of *Trionyx*

AMPHIBIA (Amphibians) [Vertebrates]

The *Amphibia,* evolved from air-breathing fish, were the first Vertebrates to leave the water to establish life on land, but had to return to water for egg-laying and the aquatic stage of their life-cycle. One of the earliest examples is the *Ichthyostega,* a primitive Amphibian, up to 3 feet in length, with Fish-like characteristics, such as the skull bones, vertebrae and tail, but Amphibian hip and shoulder girdles, ribs and limbs. Its fossil remains have been found in Upper Devonian rocks. In certain areas the Upper Devonian, Carboniferous, Permian and Triassic beds, particularly New Red Sandstone, have yielded a variety of Amphibian fossil remains. Some of these are very long and Crocodile-like in appearance, but the majority are much smaller, Lizard-size.

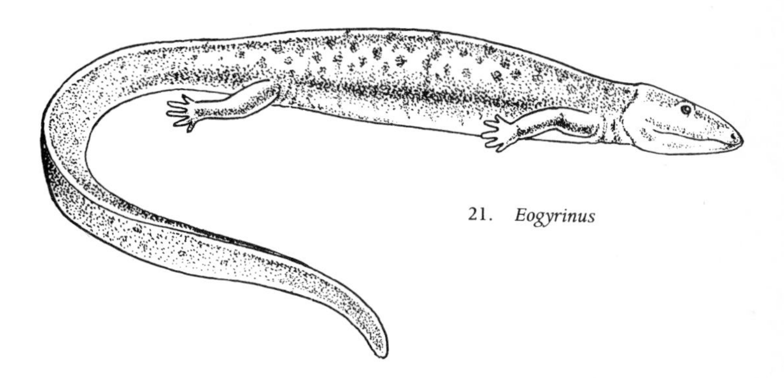

21. *Eogyrinus*

The following are a few examples. *Eogyrinus* (Fig. 21), a Salamander of the Pennsylvanian-North America, Carboniferous-Coal Measures, British Isles, up to 15 feet in length, with an almost cylindrical body, long laterally-compressed tail, weak limbs, the belly having a bony, protective shield and the head a covering of bony plates, which sometimes survive fossilized; *Eryops,* a carnivorous land Amphibian, up to 6 feet in length, fat bodied, with short limbs, of the Permian Period; *Cacops,* a small land Amphibian, up to 16 inches in length, with a large head, armour-plates along its back and a short tail, of the Permian Period; *Diplocaulus,* an aquatic Salamander, up to 2 feet in length, with a wedge-shaped head of the Permian Period.

Unlike the descendants the Reptiles, the Amphibians were not successful in their mode of life and by the close of the Triassic Period the Labyrinthodont Amphibians were extinct. Exceptions are the modern Amphibians, Newts, Salamanders, Frogs and Toads, which evolved from the Labyrinthodont Amphibians in the Carboniferous Period and were successful in their mode of life close to water, but having soft bodies and very thin bones, they did not successfully fossilize and so are rare as fossils. For this reason if they are found, the greatest care should be taken in cutting out the block to obtain the fossil Amphibian for removal. Again, like the Reptiles, sometimes only skeletal portions survive, as teeth, skull and vertebrae bones, or as moulds.

PISCES (Fish) [Vertebrates]

Bony Fishes

The *Osteichthyes* or Bony Fishes are one of the four classes of Fossil Fish. This group is also divided into two sub-classes according to their fins; the *Actinopterygii* or ray fins, and the *Choanichthyes* or lobe fins. The ray fin had a large number of supporting parallel, slender, ray-like bones, a feature of most modern Fish. The lobe fin had a structure of strong, supporting bones with smaller bones or rays and from them developed the walking limbs of the Amphibians and other Tetrapods.

Some of the primitive bony Fishes had lungs as is the case with several modern Fish genera, but the majority later had an air-bladder which controlled buoyancy. They also had large eyes and mouths essential to avoid capture yet obtain other prey. The oldest bony Fishes, in the Middle Devonian, had thick, heavy, enamelled scales, but these in later Fishes became thinner and lighter.

Of these primitive *Actinopterygians* or *Chondrostei* only a few survive as Fish today, such as the Sturgeon and the Nile Bichir.

These primitive forms were replaced, in the Triassic Period, by the *Holeost* ray fins or *Ganoids*, with more advanced skeletons, jaws and scales. Two of the present-day *Holeost* ray fins are the Garpike and Bow-fin of North American rivers, the former having been found as a fossil in Oligocene rocks in Hampshire, England, and as diamond-shaped scales in Oligocene rock, Kent, England, and the latter, Bow-fin, in the Lower Eocene beds of the Isle of Wight, but the majority of the *Holeost* ray fins were replaced by the even more advanced *Teleost* ray fins in the Jurassic and Cretaceous Periods.

An example of a fossil *Actinopterygii* or ray fin is *Belonostomus munsteri*, a fast-swimming predator with extended jaws armed with a large number of sharp, conical teeth. The body was covered with fairly thick ganoid scales. The body measured up to 18 inches in length. It has been found in the well-known German locality of Jurassic lithographic limestone at Solenhofen, Bavaria, but some examples of this genus have also been discovered in Upper Jurassic rocks in the British Isles.

The *Choanichthyes* or lobe fins are sub-divided into two orders, the Lung-fish or *Dipnoi* and the *Crossopterygians* or fringe fins.

The *Dipnoi* breathed air with nostrils that opened into the mouth to use with lungs as well as with gills, thus being able to survive in mud during drought. In Permian Period rocks in Texas, U.S.A., fossilised Lung-fish have been found in former mud burrows, with empty burrows nearby. An existing fish, the Barramunda, of Australia, has teeth and matches the fossils of Lung-fish found in Upper Palaeozoic rocks.

The *Crossopterygians* or fringe fins, so-called because the lobe fin was even more adorned, with a broad fringe of rays on the margins, were the ancestors of the Amphibians, but also include the Coelacanth, which was thought from fossils only to have existed between the Devonian and Cretaceous Periods, until examples of this 'living fossil' were caught off Madagascar in 1938 and several times since then.

The complete fossil remains of Fish are scarce and again, for this reason, if an entire or even a portion of a Fish is discovered, expert advice should be sought from the geological department of the nearest museum. Fish teeth, because these are the most durable part of a vertebrate, are more frequently found singly or in quantity. Fish vertebrae, head shields, tooth plates, spines and scales may also be found. Occasionally a 'shoal', as it were, of small Fishes are discovered together in a rock deposit. These probably buried themselves in mud during a drought and died because, for them, the lake-filling rain came too late.

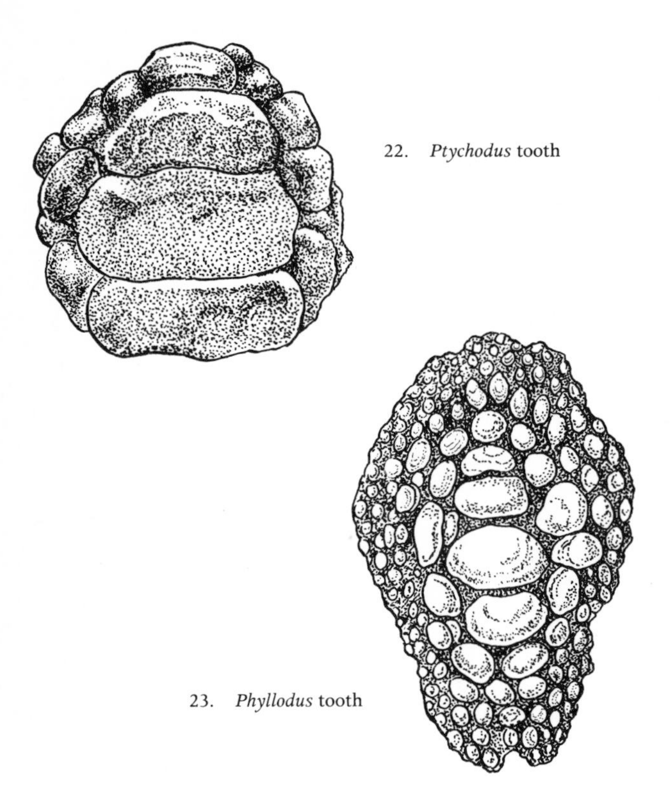

22. *Ptychodus* tooth

23. *Phyllodus* tooth

Examples of fossil Fish teeth which have been recorded are–*Asteracanthus,* tooth flattened with a pitted surface, Jurassic–Great Oolite, Forest Marble, Kimmeridge Clay; *Ptychodus,* (Fig. 22) a Ray, tooth flattened, near-cube shape with a crown of enamelled ridges, Upper Cretaceous-Chalk; *Phyllodus,* (Fig. 23) flattened, biting surface ridged, Palaeocene-Woolwich

Beds, Eocene-London Clay; *Mesodon,* teeth flattish, round, oval or kidney-bean shaped, Jurassic-Great Oolite, Forest Marble.

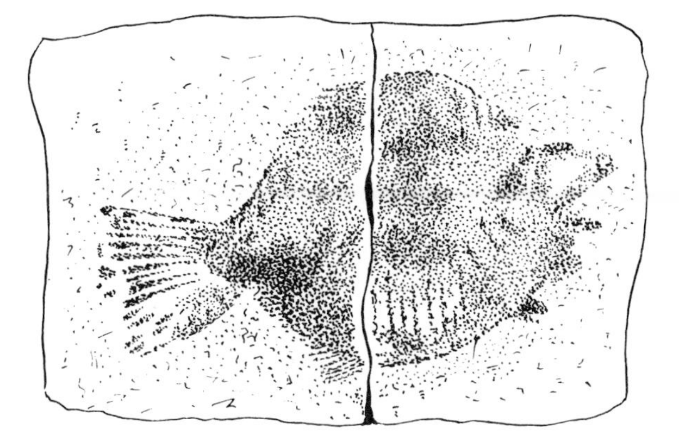

24. *Dapedium*

DAPEDIUM (Fig. 24)
Jurassic-Lias. Fossil remains entire Ganoid Fish, shaped like present-day John Dory, with deep, laterally compressed body, covered with enamelled, bony scales.

DIPTERUS
Devonian. A primitive Lung-fish. Fossil remains entire lobe fin Fish, with paired fins; body covered with thick scales, the heavy crushing teeth being sited on the mouth roof and inner side of lower jaws. Up to 15 inches in length.

LEPIDOTUS (Fig. 25)
Jurassic, Cretaceous. Fossil remains entire Ganoid, ray fin Fish, small mouthed, deep bodied, with dorsal fin set at hind part and two paired fins and anal fin underneath; body covered with heavy, enamelled scales. Up to 12 inches in length. Scales and teeth with rounded crowns also occur.

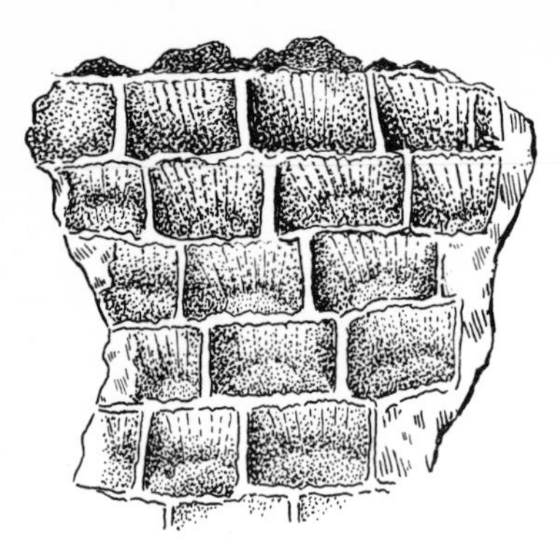

25. Scales of *Lepidotus*

OSTEOLEPIS

Devonian-Old Red Sandstone. Fossil remains entire fringe fin Fish, with well-developed, short-lobed paired fins and median fins; body covered with thick, enamelled, bony scales, the head having a shield of plate-like bones. Up to 9 inches in length.

Sharks

The *Chondrichthyes* or Sharks are the second of the four classes of fossil Fish. Their skeletons then, as now, consist chiefly of cartilage, a material which usually perished before it became fossilized. This means that complete fossil skeletons of Sharks are rare, but their enamelled teeth and spines more commonly occur in Cretaceous, Eocene, Miocene and Pleistocene rocks because the Sharks not only had large numbers in their 6 feet jaws, but these were often shed by the living Shark after use and then replaced. The oldest known Sharks occurred in the Devonian Period. Shark teeth are of two types. Those in the front of the jaws are strong and pointed to bite and grip the prey and close to the base may have pointed, curved cusps, while the teeth in the hind part of the jaws have broad flattened crowns to crush the meal. Two examples of fossil Shark teeth are *Carcharodon*, tooth broad, blunt-pointed, on almost flat base, Pleistocene-Red Crag, Eocene-Bracklesham and Barton Beds; *Odontaspis*, long, narrow pointed, curved tooth, with a deeply forked base, Eocene-London Clay (Fig. 26).

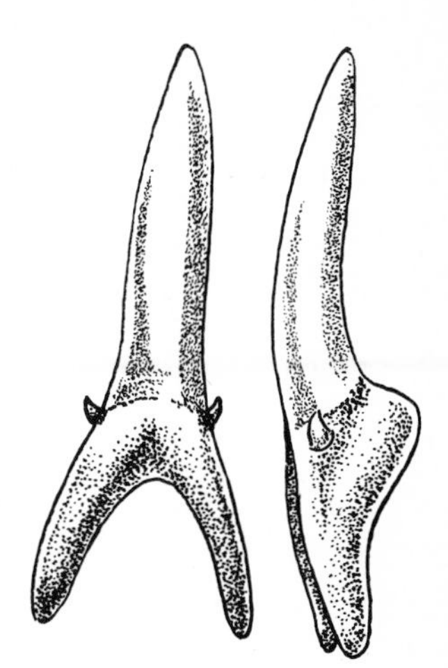

26. Tooth of *Odontaspis*

Placoderms

The *Placoderms*, 'plate-skinned', are the third of the four classes of fossil Fish and all are now extinct, their fossils occurring in the Palaeozoic Era, from the Silurian to Permian Periods. They had primitive jaws created by the modification of the anterior gill arches and also had paired fins. There were several types–the freshwater *Acanthodians*, 'spiny sharks', with several spiny fins, the streamlined body having numerous small, diamond-shaped, thick scales, the head also having some small plates; the *Arthrodires*, 'jointed necks', had an armoured head and naked body, the head-shield was hinged on a ring of thick shoulder plates, the jaws being wide; the *Antiarchs*, with a heavily-armoured head and body shield and long, jointed arm-like fins or appendages.

Entire *Placoderms* occur as fossils, also the separate plates and spines.

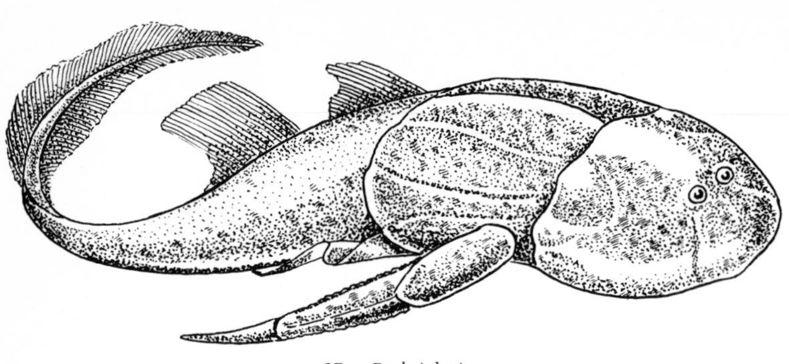

27. *Bothriolepis*

BOTHRIOLEPIS (Fig. 27)
Devonian-Old Red Sandstone. An *Antiarch*; front of body heavily armoured, with a head shield and body shield; hind part of body naked. Up to 12 inches in length.

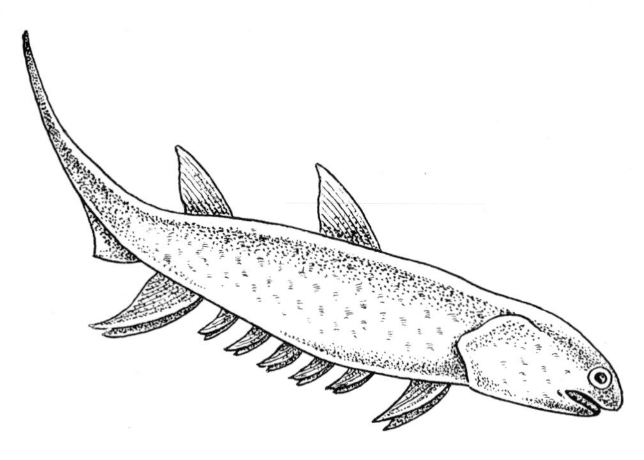

28. *Climatius*

CLIMATIUS (Fig. 28)
Upper Silurian to Devonian. An *Acanthodian*; the body covered with thick rhomboid scales and having two spines on back and five pairs of ventral fins. Up to 3 inches in length.

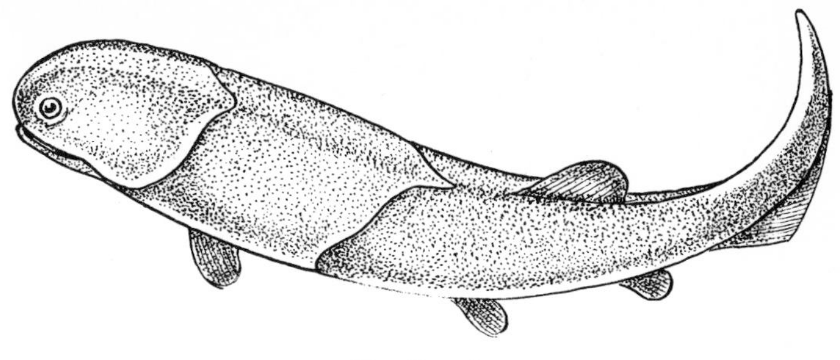

29. *Coccosteus*

COCCOSTEUS (Fig. 29)
Devonian-Middle Old Red Sandstone. An *Arthrodire*; head and thorax armoured; hind part of body naked. Up to 18 inches in length.

Agnatha

The *Agnatha*, 'jawless fish', are the fourth of the four classes of fossil Fishes and occurred very early in the evolution of the Vertebrates. These were so primitive they had no paired fins or true jaws, food being taken in through the mouth by suction. Today their living representatives are the parasitic Lampreys and Hag-fish, which attach to and feed upon their victims by sucking the latter's blood.

The majority of the fossil *Agnatha* are *Ostracoderms*, 'bony skinned', being armoured with bony plates or scales, having a thick head shield with the mouth on the underside and a powerful tail. There are two groups–the *Pteraspids* and the *Cephalaspids*. The *Pteraspids'* eyes were on the side of the head, the head shield was pointed at the front and the finless body was covered with small scales. The *Cephalaspids* had central eyes close together on top of the head focused upwards, the head shield was rounded at the front and the two-finned body covered with large, rectangular plates.

The Upper Silurian and Devonian Period rocks contain the *Agnatha* fossils; none are known from later Periods. Only rarely are complete specimens found. As a rule only the bony head shield remains because the internal skeleton was probably cartilaginous and would decay instead of being preserved.

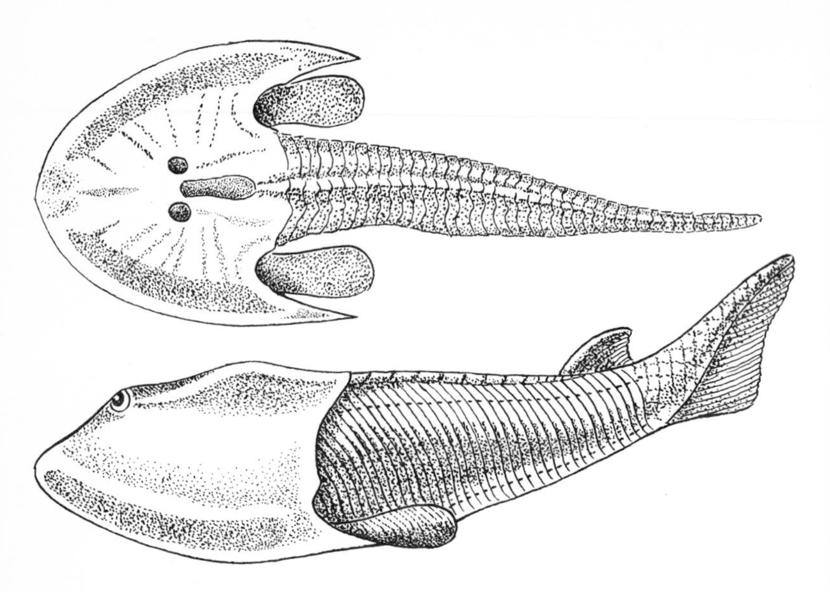

30. *Cephalaspis*

CEPHALASPIS (Fig. 30)
Upper Silurian, Devonian-Lower Old Red Sandstone. Has been discovered as an entire fossil in Angus, Scotland. It had a flattened, bony armoured head shield, rounded anteriorly, with the trunk and tail plate-covered. Up to 12 inches in length.

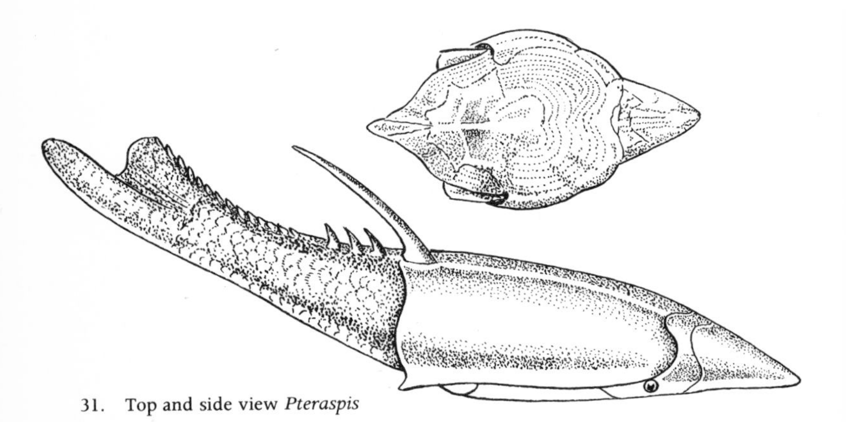

31. Top and side view *Pteraspis*

PTERASPIS (Fig. 31)
Upper Silurian, Devonian-Lower Old Red Sandstone. It had a large head shield pointed anteriorly, with a protective spine on the rear upper edge of this, the rear body and tail being Fish-like with small scales. Up to 6 inches in length.

THELODUS
Middle Silurian-Llandovery, Lower Devonian-Lower Old Red Sandstone. It was Tadpole-shaped with all the flattened body covered with tiny, collar-stud like, interlocking, bony, enamelled scales. Three to 8 inches in length.

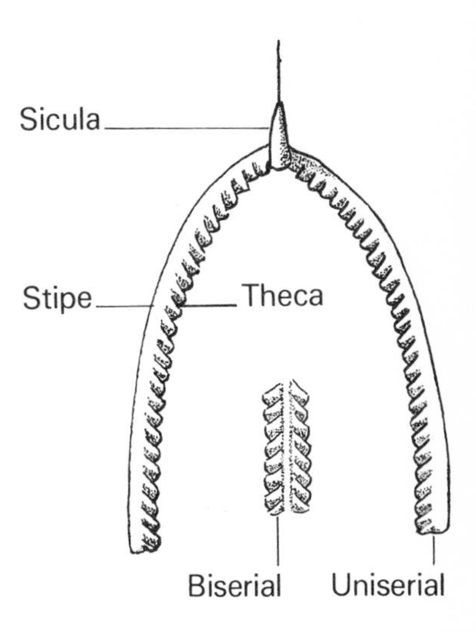

32. Structure of Graptolite

GRAPTOLITES

(Fig. 32)
The *Graptolites* were small, colonial, marine organisms related to the *Pterobranchs*, which in turn belong to the *Chordata* and so were closely related to the *Vertebrates*, For this reason I have included them at the end of the *Vertebrate* Fossil Examples Chapter and not at the beginning of the Invertebrate Fossil Examples Chapter. A *Graptolite* consisted of one or several chitinous branches, stems or 'stipes' bearing small cup-like

structures or 'thecae'. These 'thecae' bearing 'stipes' grew in various forms. Some were numerously branched, others were in groups of three or four branches, or were single, while some examples resembled a tuning-fork in shape. They commonly occur either as white films or pyritized rods, in shales, from the late Cambrian to the early Carboniferous (British Isles), Mississippian (North America) Periods, but in the British Isles became extinct in the Silurian Period.

DENDROIDS

Upper Cambrian to Carboniferous (British Isles), Mississippian (North America) Periods. Much branched early *Graptolites* with thin stipes. May be fan-shaped or bushy with a root-like base. There are several forms, one being *Dictyonema,* which occurs in Cambrian-Tremadoc Series, in Britain and North America. It has numerous branches or stipes joined with short bars so the stipes hang down in a pendant pyramid-like form from the sicula, the latter being the first thecal cup that founded the colony. Up to 4 inches in length.

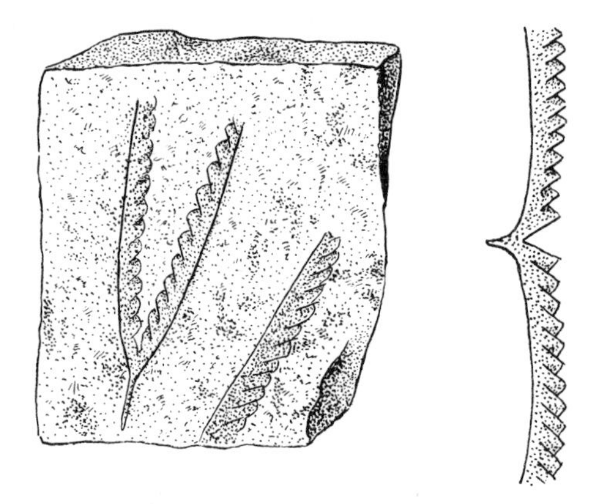

33. *Didymograptus*

DIDYMOGRAPTUS (Fig. 33)

Lower and Middle Ordovician-Llanvirn, Arenig, Llandeilo Series. It occurred in two forms. *D. bifidus* had two stipes, each with one row of cylindrical thecae, connected at one end in a rod or nema to create a pendent or tuning-fork prong shape. Stipe length 1 to $2\frac{1}{2}$ inches. *D. extensus* also had two stipes but these were horizontal, as if the *Graptolite* had been bent to appear almost straight.

DIPLOGRAPTUS (Fig. 34)
Middle Ordovician-Caradoc and Llanvirn Series, Lower Silurian-Llandovery Series. Comprises a single stipe with a row of close, oblique, curved thecae on each side. About 2 inches in length.

34. *Diplograptus*

MONOGRAPTUS
Silurian-Llandovery, Ludlow Series. A single straight or slightly curved stipe with a single row of thecae on one side. There are several forms of *Monograptus* identified by the thecae shape. Up to 2 inches in length.

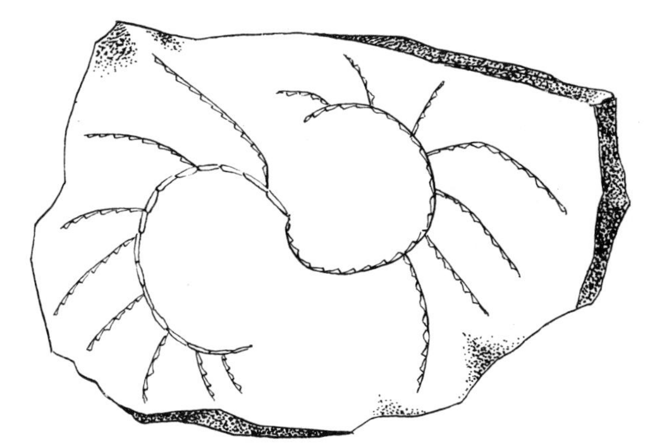

NEMAGRAPTUS (Fig. 35)

Middle Ordovician-Llandeilo, Caradoc Series. An easily identified fossil *Graptolite* because the two stipes form an S-shape from which shorter branches radiate. In some examples the 'S' may be obvious, in other examples this curving shape is less defined. Up to 1½ inches in length.

FOSSIL EXAMPLES
Part Two
Invertebrates

ARTHROPODS [Invertebrates]

The *Arthropods*, 'joint-footed', are a large group of *Invertebrates* with segmented bodies that have a rigid, hard, protective, chitinous, external covering over their soft parts; paired, jointed appendages or limbs; antennae; mandibles, and, depending on the class, wings. They are divided into five classes–*Trilobitomorpha, Myriapoda, Crustacea, Chelicerata, Insecta*. The *Myriapoda* (Centipedes and Millipedes), are very rare as fossils in the Palaeozoic Era, and thus unlikely to be discovered so I have not elaborated on them. An example is *Acantherpestes,* Pennsylvanian, North America, an 8 inch long Millipede with short legs, similar to modern Millipedes; two other Millipedes being *Euphorberia* and *Xylobius* in Carboniferous rocks in the British Isles.

Trilobitomorpha

This class contains the *Trilobites,* marine *Arthropods* which first occurred in the Lower Cambrian Period and became extinct in the Permian Period. They vary considerably, with three major divisions of the body–the cephalon or head shield, thorax, and pygidium or tail piece–the thorax comprising three-lobed segments, hence their name. Fig. 36 indicates the positions of the anatomy referred to in the description of examples. The parts usually commonly fossilized are the hard, chitinous dorsal surface or back. *Trilobite* fossils may occur flat or rolled into a near-ball, like the present-day Woodlouse, Sow or 'Pill-Bug', or distorted by sediment movement and pressure during fossilisation, or broken into fragments. During their growth *Trilobites* moulted their chitinous skeleton and the various size stages of these, sometimes in portions, can be found.

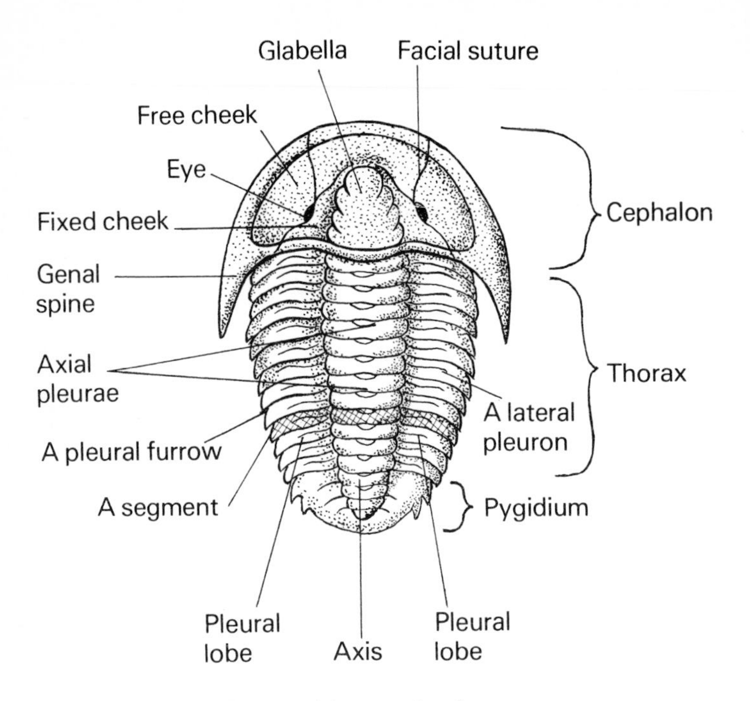

36. Trilobite Dorsal Surface

AGNOSTUS (Fig. 37)
Middle Cambrian. The cephalon and pygidium are the same size so the small two-segmented thorax has a 'pinched-in waist' appearance. Blind. Abundant in Cambrian, but very small, about $\frac{1}{4}$ inch in length.

BATHYURISCUS
Middle Cambrian-West North America. Has up to 9 thorax segments; the cephalon is semi-circular with two short genal spines; the glabella is furrowed, the pygidium being segmented and semi-circular in shape. Eyes crescent-shaped. Up to $1\frac{1}{2}$ inches in length.

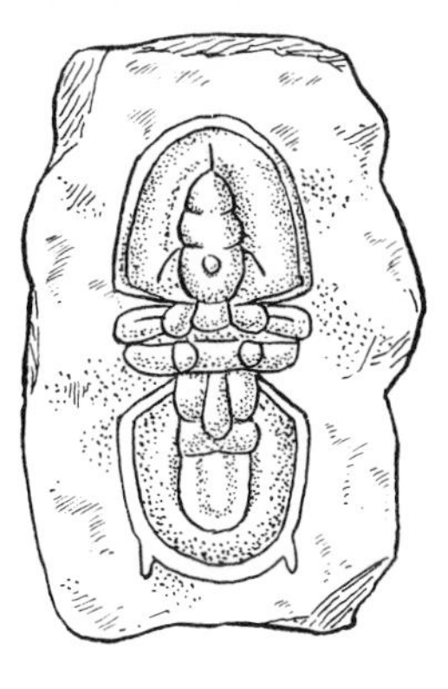

37. *Agnostus*

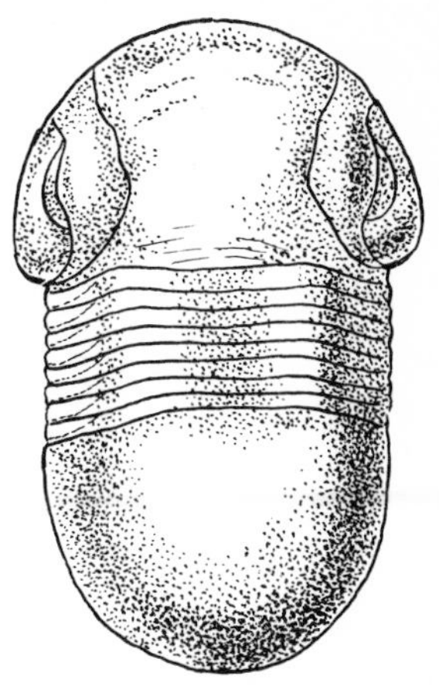

38. *Bumastus*

BUMASTUS (Fig. 38)
Ordovician to Silurian. Cephalon and pygidium globular without segments, the entire body having an oval shape, with a wide axial lobe; no genal spines. Eyes kidney-shaped. About 4 inches in length.

CALLAVIA
Lower Cambrian. Cephalon semi-circular, with short genal spines, the pygidium being minute and almost non-existent, entire body oval-shaped. The glabella is long and narrow with a long spine where cephalon joins the thorax. Eyes crescent-shaped. Up to 6 inches in length.

CALYMENE (Fig. 39)
Silurian-Wenlock Limestone, Devonian. Cephalon semi-circular, but no genal spines, pygidium also semi-circular, the glabella being narrow and divided into lobules by deep grooves. There are 13 grooved thorax segments. From $1\frac{1}{2}$ to 3 inches in length.

DIKELOCEPHALUS
Upper Cambrian. Cephalon broad; glabella short, blunt and furrowed; the thorax has 12 segments, the pygidium being broad with two spines. Up to 6 inches in length.

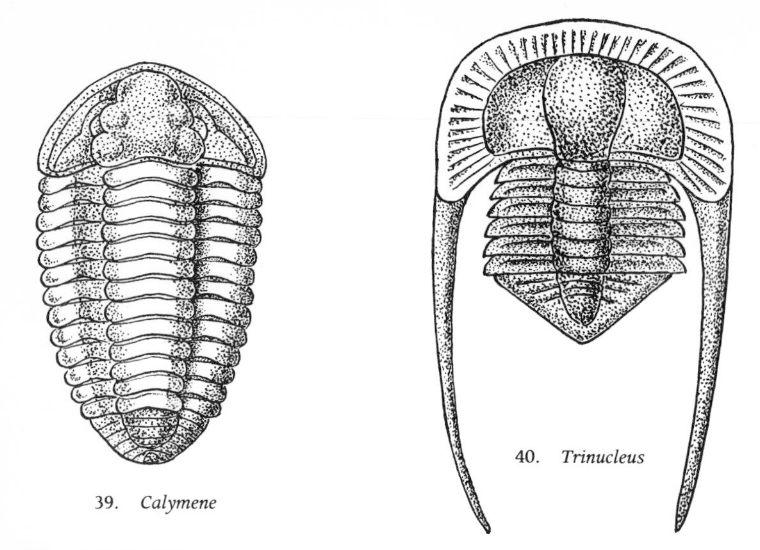

39. *Calymene*

40. *Trinucleus*

FLEXICALYMENE

Ordovician-Ashgill, Llandeilo, Silurian. Very similar to *Calymene*, but has a lip-like margin in front of the glabella. Up to 2 inches in length.

GRIFFITHIDES

Mississippian (North America), Carboniferous (British Isles). Cephalon semi-circular; the glabella wide and furrowed; pygidium also semi-circular with up to 16 segments, the thorax having 9 segments. The entire body is oval shaped. Eyes small. Up to 2 inches in length.

PARADOXIDES

Middle Cambrian. Cephalon semi-circular with long genal spines; glabella expanding forward; also numerous, spined thorax segments, but pygidium is small. Up to 18 inches in length.

TRINUCLEUS (Fig. 40)

Middle Ordovician-Llandeilo, Llanvirn, Caradoc. Cephalon large, broad and pitted, with very long genal spines; the thorax and pygidium being shorter in size than the cephalon; glabella inflated. Blind. Up to 1 inch in length.

Crustacea [Invertebrates]

The *Crustacea* are chiefly aquatic, marine and freshwater, *Arthropods,* with two pairs of antennae and several pairs of two-branched appendages, occurring in the Cambrian Period up to and including the Recent Period. *Crustaceans* occurring as fossils include the *Ostracods, Branchiopods, Cirripedes* (Barnacles) and *Malacostracous Crustaceans* (Shrimps, Crabs, Lobsters).

OSTRACODS are small *Crustaceans,* up to 1 inch in length, with the body enclosed in a bivalved, laterally compressed, calcareous shell, hinged on its upper margin. Sometimes one shell valve overlaps the other and the shells may be pitted, lobed, ribbed or spinous. They occur from the Cambrian Period to Recent Period, in shales, marls, limestones and the small examples, requiring microscopic examination for identification, may be extremely numerous.

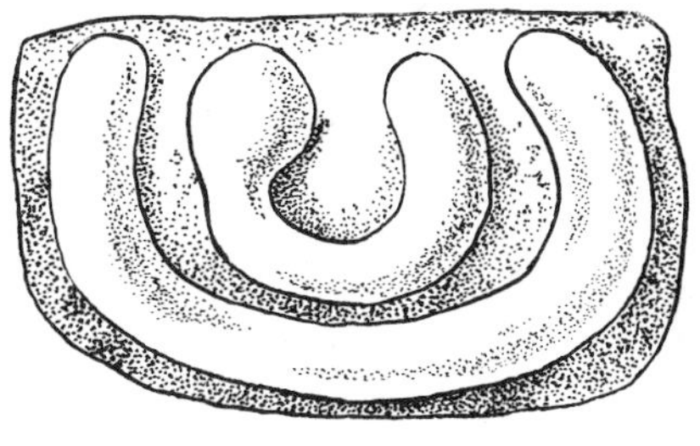

41 *Bollia*

BOLLIA (Fig. 41)

Ordovician to Devonian. A swollen, long outer ridge has a parallel horseshoe-shaped ridge within it, the shell also having a pitted surface. Shape, sub-oval. Up to $\frac{1}{4}$ inch in length.

CYTHEREIS

Cretaceous. Ridges vary in shape and amount–curved, straight or irregular. Shape, anterior end is deep, rounded and notched, top and bottom margins straight, hind end being compressed, narrow, rounded with a few notches. Valves are unequal. Up to $\frac{1}{3}$ inch in length.

DIZYGOPLEURA

Silurian to Devonian. Not to be confused with *Bollia*. It has four, prominent, vertical lobes with the inner lobes continuous ventrally. Shape, sub-oval. Valves are unequal, the left valve overlapping the right valve. Up to $\frac{1}{4}$ inch in length.

GLYPTOPLEURA

Mississippian (North America), Carboniferous (British Isles) to Permian. Has several prominent horizontal ridges. Shape, sub-oval with a straight hinge. Valves are unequal, the right valve overlapping the left valve. Up to $\frac{1}{4}$ inch in length.

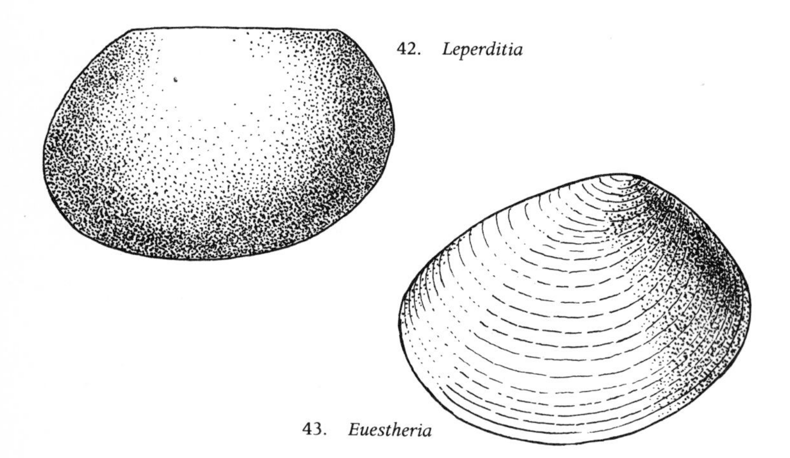

42. *Leperditia*

43. *Euestheria*

LEPERDITIA (Fig. 42)

Lower Silurian-Wenlock, Ludlow to Upper Devonian. Plain but finely pitted. Shape, oblong with short, straight hinge. Valves unequal. Up to $\frac{1}{4}$ inch in length.

BRANCHIOPODS, not to be confused with the *Brachiopods,* have a bivalved, chitinous carapace, but without a hinge. In the Rhaetic and Upper Keuper beds of the Triassic Period the carapaces of *Euestheria,* similar to a very small Lamellibranch shell, and *Pseudomonotis,* similar to a very small Cockle, occur, the latter also occurring in the Magnesian Limestone series of the Permian Period (Fig. 43).

CIRRIPEDES (Barnacles) are common now as seashore creatures but less so as fossils, although they date from the Eocene Period to Recent Period. They comprise several calcareous, triangular plates forming an approximate pyramid-shape and these may be found separately as fossils. Acorn Barnacles (Balanus) are more likely to be found as fossils upon Mollusc shells wherever these occur, such as the Pliocene Period-Coralline Crag Series and Pleistocene Period-Red Crag Series.

MALACOSTRACOUS CRUSTACEANS (Shrimps, Crabs, Lobsters) are also common now as marine creatures, but occur less often as fossils, although they date from the Cambrian Period to Recent Period. Their anatomy is too familiar to require description and fossil Crabs, Lobsters, Shrimps, are fairly easy to recognise. The carapace of the Crabs is similar to that of today's species; the limbs may either be absent or tucked underneath and fossilized in that position. Crabs frequently occur in the Eocene Period-London Clay Series.

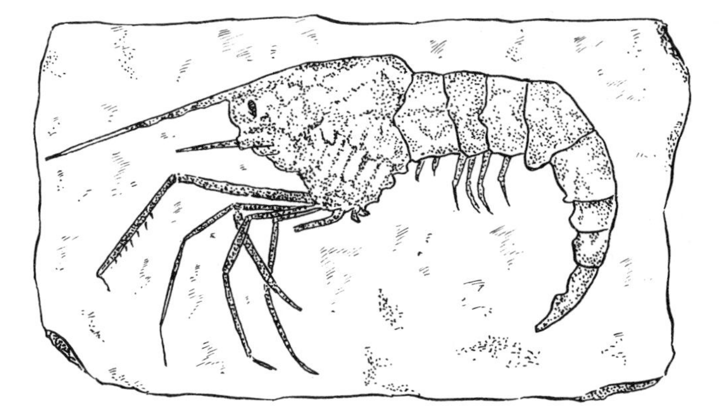

44. *Aeger*

AEGER (Fig. 44)
Jurassic. Typical Shrimp body, laterally compressed, with long limbs. Up to 6 inches in length.

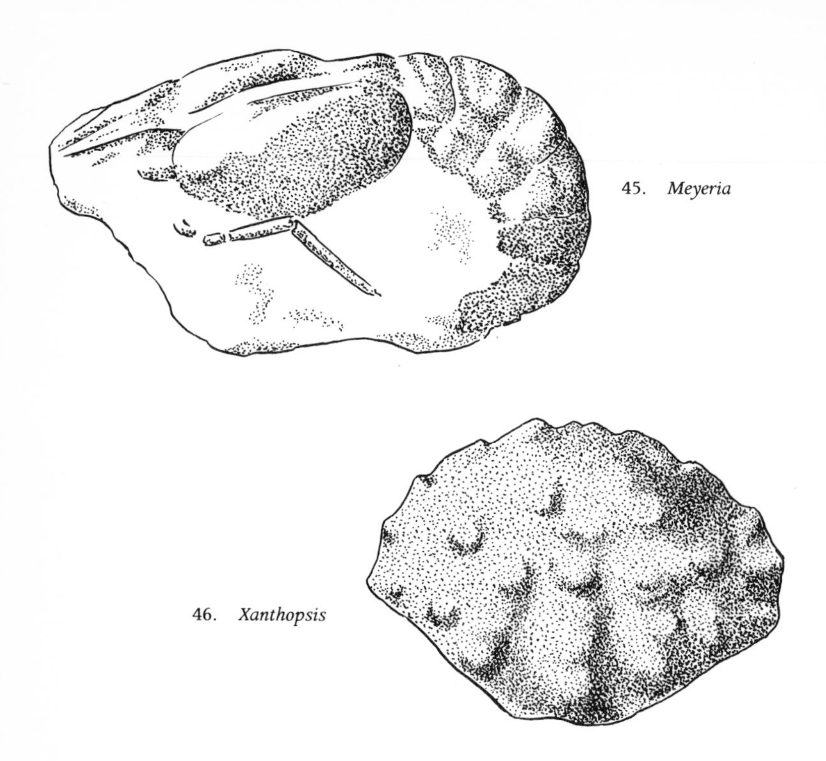

45. *Meyeria*

46. *Xanthopsis*

ERYON
Jurassic. A *Decapod* similar to a Lobster in shape, with a wide carapace. Up to 4 inches in length.

MEYERIA (Fig. 45)
Cretaceous-Lower Greensand. A Lobster. Up to 6 inches in length.

XANTHOPSIS (Fig. 46)
Eocene-London Clay. A Crab. Up to $2\frac{1}{2}$ inches in width.

Chelicerata [Invertebrates]

The *Chelicerata* have the head and thorax fused to form the cephalothorax and in some classes a pair of the anterior limbs have pincers. This group of *Arthropods* is divided into the *Arachnids* (Spiders, Scorpions, Mites, etc.); *Merostomes*, and *Eurypterids*.

ARACHNIDS

Fossil Spiders occur from the Devonian Period and fossil Scorpions from the Silurian Period, but both are rare as fossils. In the Carboniferous rocks in Britain a primitive fossil Spider, *Protolycosa,* has been obtained; and *Architarbus,* a Spider, up to 4 inches in length, with a broad cephalothorax and short abdomen, has been found in Pennsylvanian rocks in the U.S.A. In the Upper Silurian rocks a typical-shaped Scorpion, *Palaeophonus,* up to 2 inches in length, with pincer limbs, which also had gills and lived in the sea, has been found (Fig. 47). In the Upper Triassic rocks numerous fragments of another Scorpion, *Mesophonus,* were also discovered.

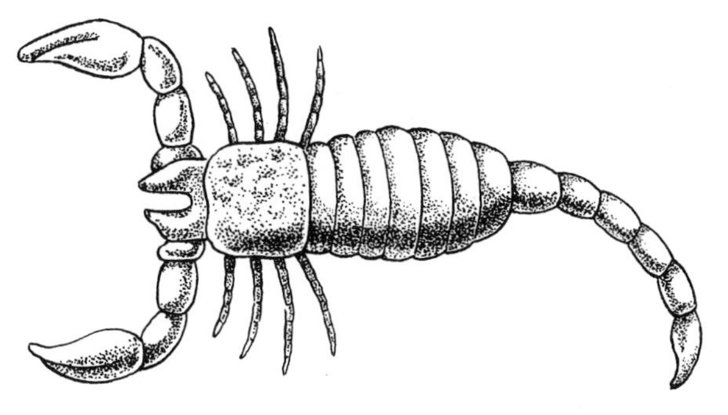

47. *Palaeophonus*

MEROSTOMES

The *Merostomes* occasionally occur as fossils from the Cambrian Period. One of these is *Limulus,* similar to the present day King Crab, with a rounded body and long tail spine.

EURYPTERIDS

The extinct *Eurypterids* or 'Sea Scorpions', had a flattened, broad-fronted, segmented, chitinous body which tapered to a spike-like tail. The cephalothorax had two pairs of eyes and six pairs of limbs, the front pair long and bearing pincers and hind pair oar-like. They were aquatic, probably mainly carniverous, feeding on smaller Invertebrates seized with the pincers. They occur in beds from the Ordovician Period to the

Permian Period, but are chiefly found in the Silurian Period and Devonian Period beds. Some examples were up to 10 feet in length and powerful swimmers in pursuit of prey (Fig. 48). *Pterygotus*, Devonian-Upper Old Red Sandstone, reached 10 feet in length, but the average is 12 inches. *Stylonurus*, Silurian to Devonian, 8 inches in length, had very long limbs, a narrow, tapered body and long, pointed tail-spike. *Hughmilleria,* Ordovician to Permian, up to 3 inches in length, had a semi-circular head, short limbs, except for the hind pair which are longer and armed with tiny pincers, and a blade-shaped tail.

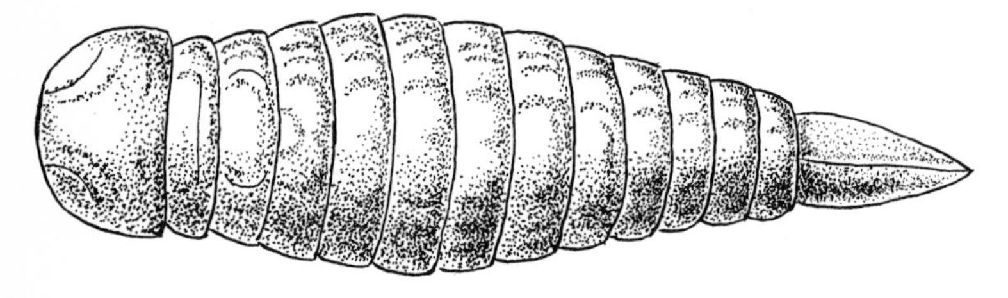

48. *Pterygotus*

Insecta (Insects) [Invertebrates]

The *Insecta* have a three-part body comprising a head, thorax and abdomen, but the varied forms are too numerous to be described here. Several hundred fossil Insect forms have been discovered, but, because of their frailness, they are rare. The earliest known are wingless Insects occurring in the Devonian Period-Middle Old Red Sandstone. In the Carboniferous Period (British Isles), Pennsylvanian Period (North America) some of the Insect examples were gigantic compared with their modern counterparts. Other forms, however, have changed very little in shape and construction from these ancient species. Peat contains broken fragments and occasionally complete specimens of Beetles and other early swamp Insects. In the Coal Measures the Insects may be fossilized only as impressions of the wings and body. These Coal Measures often contain Ironstone nodules (See Chapter One) which have fossils inside them, such as

Insects, Fish, etc. Lake deposits, such as the Miocene example at Florissant, Colorado, U.S.A., contain Insects. Many Insects have been found trapped in amber of the Oligocene Period. Most familiar of the fossil Insects are the forerunners of the Dragonflies which they resemble (Fig. 49), and occurred from the Permian Period to Recent Period. *Tarsophlebia* occurred in the Jurassic Period, but only had a wingspan of 2 inches. There were other Dragonflies with a wingspan of 2 feet. *Meganeura,* in the Pennsylvanian Period, had a wingspan of 30 inches.

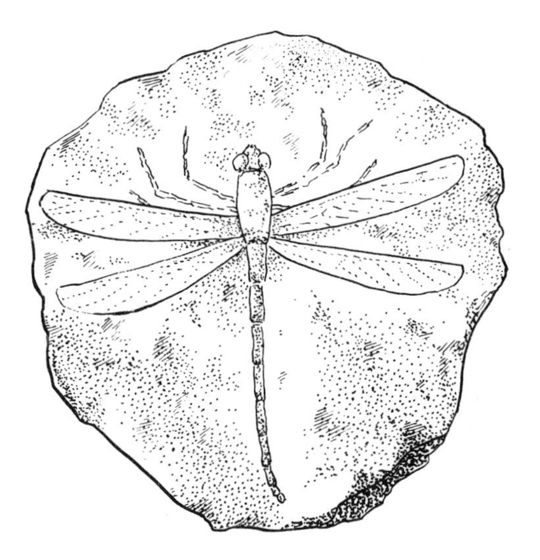

49. Dragonfly—*Tarsophlebia*

Blattoids, not to be confused with the *Blastoids,* were an ancient group of Cockroaches, up to 4 inches in length occurring from the Pennsylvanian Period to the Recent Period. From the Permian Period to the Recent, Lacewing Flies (Neuroptera) have occurred, their name due to the frail lace-like transparency of their wings. *Mesopsychopsis* is an example, up to 1 inch in length, of the Jurassic Period. Other fossilized Insects are Mayflies and Grasshoppers.

ECHINODERMATA

The *Echinodermata*, 'spiny-skinned animals', are a group of marine animals with bodies comprising limy plates or spines. They are sub-divided into two groups–the *Pelmatozoa*, comprising the classes *Cystoidea, Blastoidea, Crinoidea,* and *Edrioasteroidea,* all of which existed attached to their habitat, and the *Eleutherozoa,* comprising the *Echinoidea, Stelleroidea,* and *Holothuroidea,* all of which were and are free moving in the habitat.

Pelmatozoa

CYSTOIDS, now extinct, were a class of *Echinoderms* with a primitive rounded sac-like body, comprising numerous enclosing, protective, irregularly positioned, calcareous plates with arranged, triangular pore openings and grooves linked to the mouth on the upper surface for food collection, there being no long sinuous arms for this purpose, as in the *Edrioasteroids.* The body was attached at its base to a long stem. *Cystoids* occurred from the Ordovician Period to the Devonian Period.

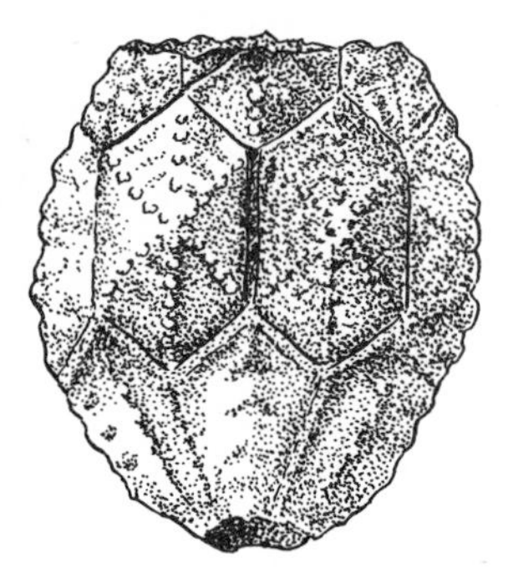

50. *Caryocrinites*

CARYOCRINITES (Fig. 50)
Ordovician and Silurian. Body globular-shaped with large, regularly-positioned plates. May have remains of its up to 13 weak arms and long stem. Up to 1 inch in diameter.

Plate V *Ophioderma egertoni* (Broderip). Marlstone, Middle Lias, Jurassic. Lyme Regis, England.

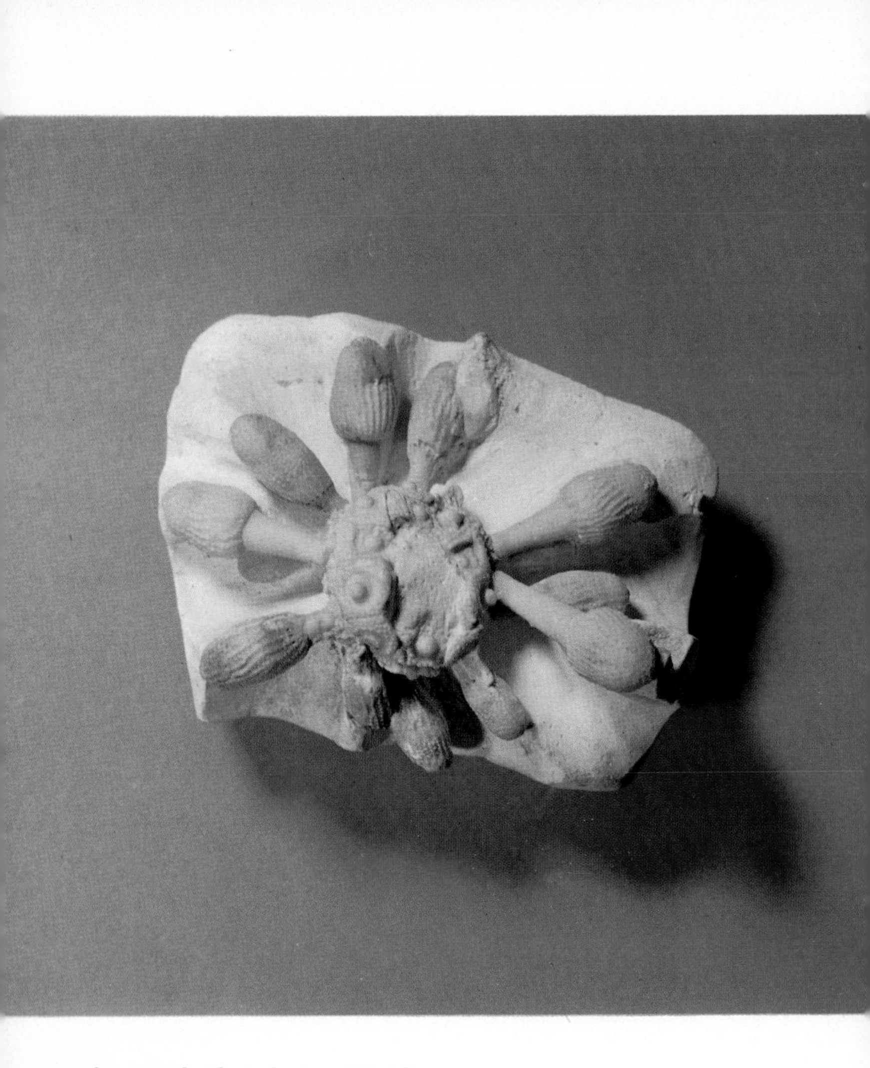

Plate VI *Tylocidaris clavigera* (König). Upper Chalk, Cretaceous. Gravesend, Kent, England.

ECHINOSPHAERITES

Ordovician. Body globular-shaped with numerous irregular polygonal plates that have obscure, indefinite borders. Long stem. Up to 2 inches in diameter.

BLASTOIDS, now extinct, were also a class of *Echinoderms,* with a cup-shaped body, comprising a number of symmetrically positioned plates with five petal-shaped ambulacral grooves that bore the numerous brachioles for food collection. The body was attached at its base to a short stem. *Blastoids* occurred from the Ordovician Period to the Permian Period, but are not common as fossils, those in the Mississippian Period (North America), Carboniferous Period (British Isles) being most frequently found.

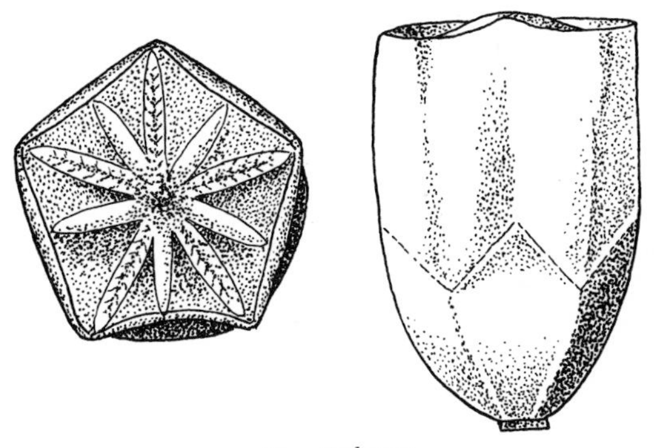

51. *Codaster*

CODASTER (Fig. 51)

Silurian, Devonian, Mississippian (North America) Pennsylvanian (North America), Lower Carboniferous (British Isles). Body pyramidal cup-shaped, tapered at its base, ambulacral grooves being short, triangular. Up to 1 inch in height.

CRYPTOBLASTUS

Mississippian (North America). Body globular cup-shaped with lengthy ambulacral grooves, side plates are large, overlapping those at the mouth. Up to $\frac{1}{2}$ inch in height.

PENTREMITES (Fig. 52)
Mississippian, Pennsylvanian (North America). Small body a bud-like cup-shape with lengthy side plates around the broad, petal-like ambulacral grooves, the basal plates being small. Up to 2 inches in height.

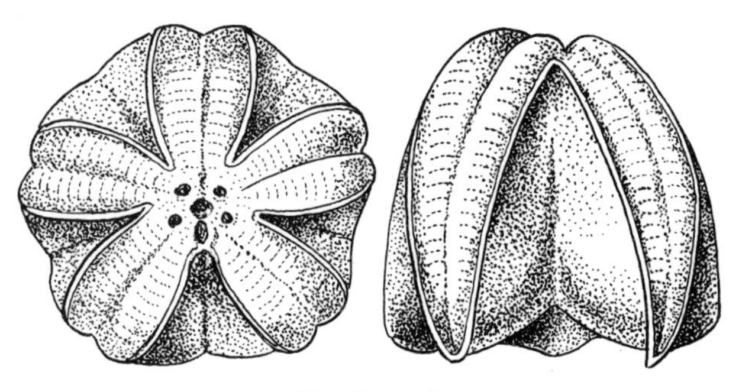

52. *Pentremites*

(Fig. 53)
CRINOIDS exist today as 'Sea Lilies' in varying sea depths, this name being derived from their similarity in shape to a flowering plant. Their structure consists of a cup, at the top of a long, slender, flexible, calcareous stem, consisting of circular discs or variously shaped columnals, anchored to the habitat by a root-like structure or a cement. The cup consists of five radial plates around the top of which are food-collecting arms, each arm having a groove. These arms bear numerous vibrating hairs which create a current to convey microscopic food, animals and plants, down the groove to the Sea Lily's mouth. In this way it collects and secretes the lime to add to its stem and skeleton. Most of the Sea Lilies usually live a sedentary existence but can if conditions demand, break away from the stem to seek a better habitat. There are also pelagic examples.

Fossil *Crinoids* occur from the Ordovician Period to the Recent Period but were particularly varied and numerous during the Silurian, Devonian and Carboniferous Periods (British Isles), Mississippian Period (North America). They may very rarely be found intact, with arms, cup and the stem, but fragments of these are much more frequently discovered. There were also free-swimming, stemless *Crinoids*, which occur as fossils in the Cretaceous Period.

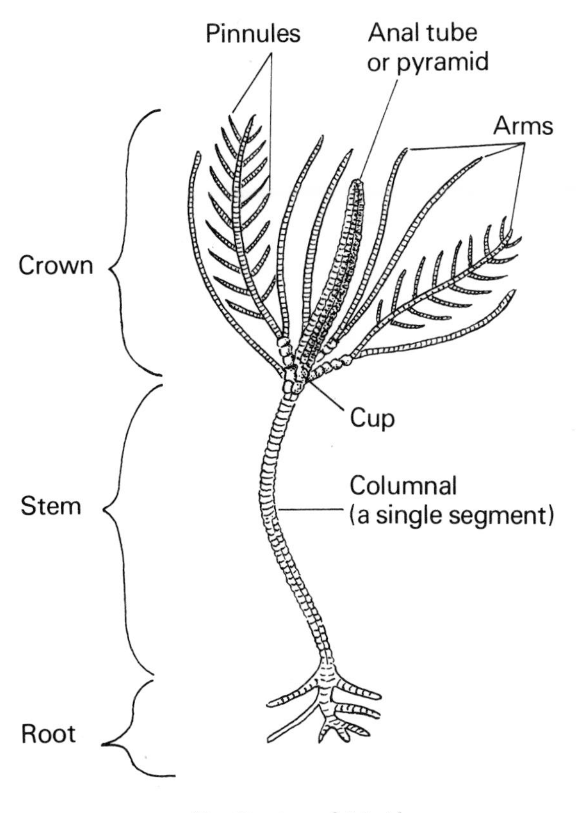

53. Structure of *Crinoid*

APIOCRINUS (Fig. 54)
Middle Jurassic. Long cylindrical stem formed of bead-like discs which increase in size at the top of the stem where they join the cup's basal plates to form a pear shape; five branched arms. Height of crown about 2 inches.

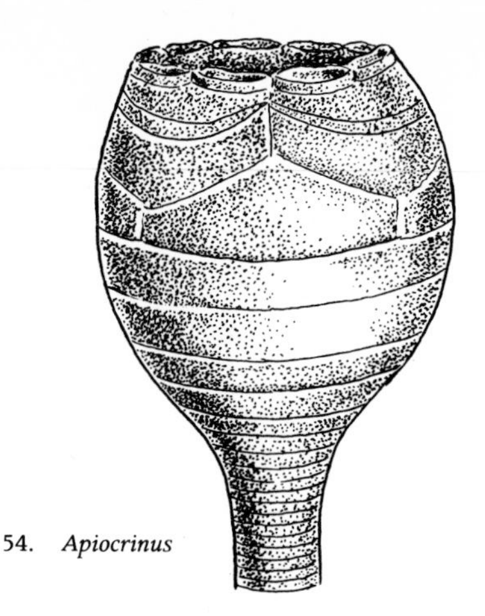
54. *Apiocrinus*

GLYPTOCRINUS
Ordovician, Silurian. Cup small; arms long, narrow and branched; has star-shaped patterning. Height of crown about $2\frac{1}{2}$ inches.

ISOCRINUS
Triassic. Long stem formed of star-shaped columnals; cup small; five long, branched arms forming a large crown. Similar to *Pentacrinus*. Height of crown up to $2\frac{1}{2}$ inches.

PENTACRINUS (Fig. 55)
Lower Jurassic-Lower Lias. Long stem formed of star-shaped columnals; cup small; five long, branched arms forming a large crown. Height of crown about $2\frac{1}{2}$ inches.

PLATYCRINUS
Mississippian (North America), Pennsylvanian (North America), Carboniferous (British Isles), Permian. Long stem formed of discs; cup deep, may be rough-surfaced; arms long and branched. Height of crown up to $2\frac{3}{4}$ inches.

TAXOCRINUS (Fig. 56)
Devonian, Carboniferous (British Isles), Mississippian (North America). Long stem formed of discs; cup small; very large branched arms. Height of crown about 2 inches.

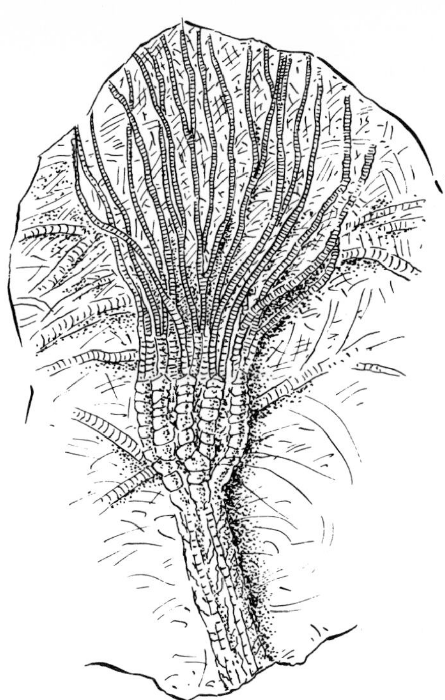

55. *Pentacrinus*

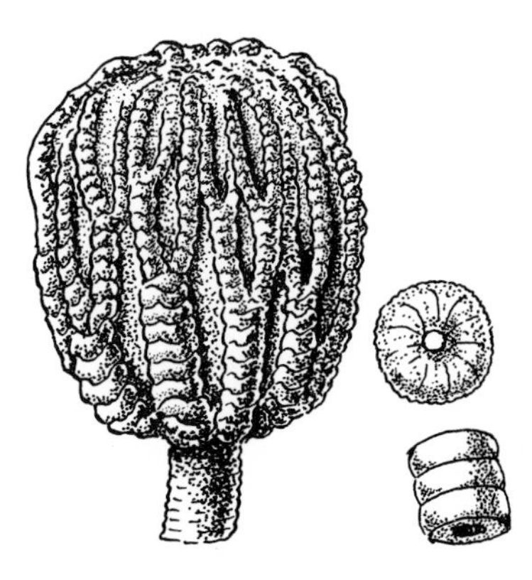

56. *Taxocrinus*

EDRIOASTEROIDS, now extinct, were a class of *Echinoderms,* attached to the habitat, with a rounded or flattened, sac-like body comprising numerous small, flexible, irregularly-positioned polygonal plates. The central mouth was surrounded by five, slender, curved, unbranched, armlike, ambulacral areas. *Edrioasteroids* occurred from the Cambrian to Mississippian (North America), Carboniferous (British Isles) Periods, but are very rare as fossils.

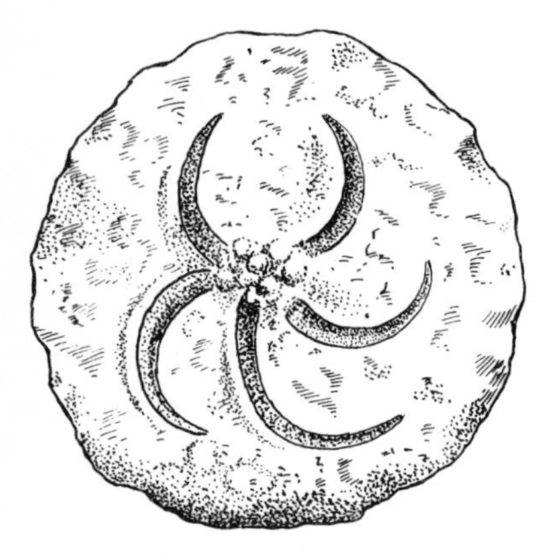

57. *Agelacrinites*

AGELACRINITES (Fig. 57)
Devonian, Carboniferous (British Isles), Mississippian (North America). Has five slender, curved, ambulacral grooves, two positioned curving to the left side, three positioned curving to the right side; around the edge is a circular pattern of small plates. Up to 1½ inches in diameter.

HEMICYSTITES
Ordovician, Silurian, Devonian. Has five short arms radiating starfish-like, body being thin and flattened, with encircling rings of large and small plates. Up to ¾ inch in diameter.

Eleutherozoa

STELLEROIDS (British Isles), ASTEROIDS (North America)

exist today as the familiar free-moving Starfish (Echinoderms), with five thick, broad 'arms' radiating from the central disc body and strengthened by large numbers of limy plates. The mouth is on the centre underside of the disc. Wide ambulacral grooves on the lower surface of the 'arms' bear numerous tube-feet ending in suckers used by the Starfish to crawl and climb. *Stelleroids/ Asteroids* occur as fossils from the Ordovician to Recent Period, but are uncommon. They are more likely to be found in fragments as the body tended to disintegrate and decompose after death and for this reason entire fossil specimens are rare.

DEVONASTER [MESOPALAEASTER] (Fig. 58)
Ordovician, Devonian. The arms are noticeably pointed and the ambulacral grooves are prominent, the body disc and arms having numerous small plates. Up to $1\frac{1}{2}$ inches in diameter.

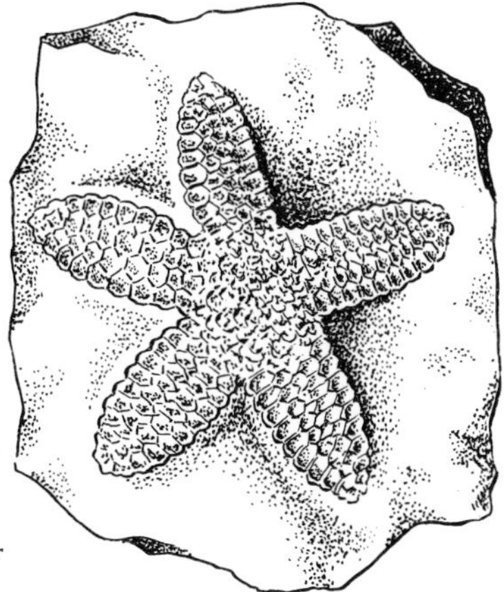

58. *Devonaster*

HUDSONASTER
Ordovician. The arms are short and thick, similar to present-day Cushion Starfish, the body and arms having a covering of large plates. Up to 1 inch in diameter.

URASTERELLA

Ordovician to Pennsylvanian (North America), Carboniferous (British Isles). The arms are long, narrow, having prominent ambulacral grooves and the plates are small and irregular, but the body disc is minute with no clear demarcation between the central disc and arms. Up to 2 inches in diameter.

OPHIUROIDS exist today as the familiar free-moving *Echinoderm* relatives of the Starfish, the Brittle Stars, so-named because they readily dispose of one or more of their five flexible arms if trapped or seized by an enemy. The central body disc is usually clearly shaped and small in comparison with the long, slender arms. Instead of suckers on their tube-feet they progress by writhing, jerking movements of their arms. *Ophiuroids* occur as fossils from the Ordovician Period to Recent Period, but, like the *Asteroids/Stelleroids* they are uncommon, entire Brittle Star fossils being rarer than fragmentary remains.

AGANASTER

Mississippian (North America), Carboniferous (British Isles). The narrow arms taper from a flower-shaped central body disc. Up to 1 inch in diameter.

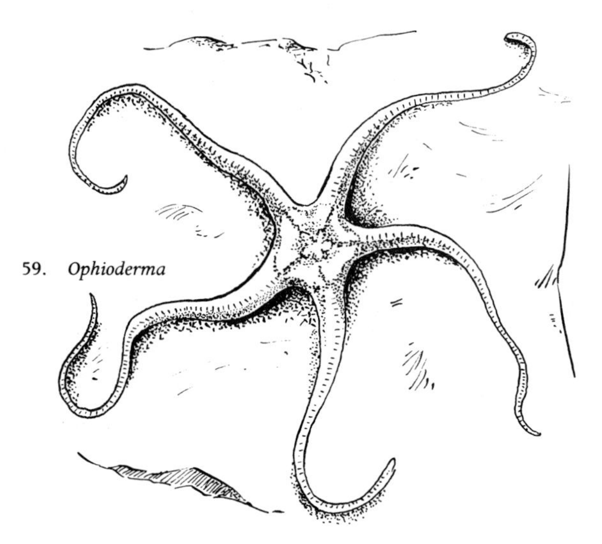

59. *Ophioderma*

OPHIODERMA (Fig. 59)
Jurassic-Lias. The arms are typically long and slender, tapering from a rounded petal-like body disc. Up to $1\frac{1}{2}$ inches in diameter.

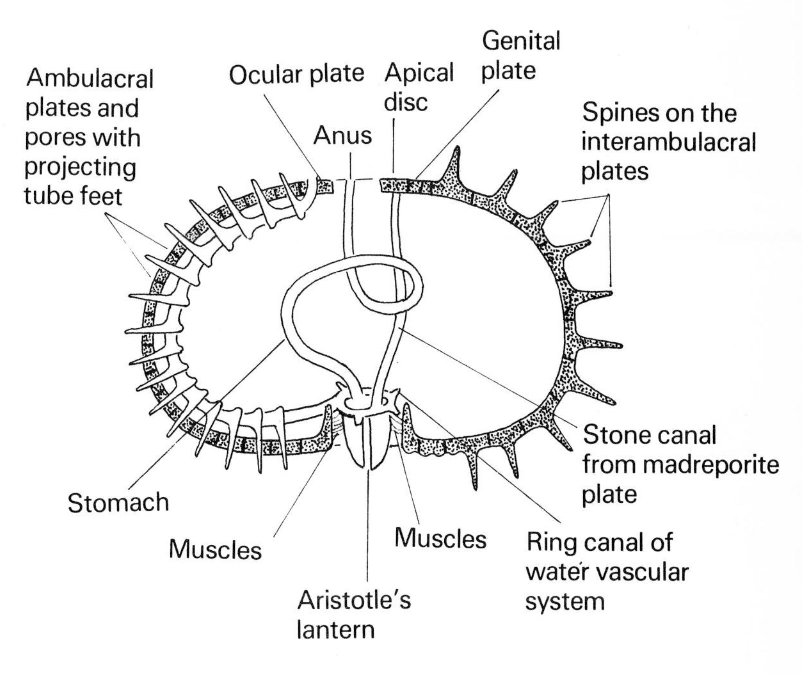

60. Structure of *Echinoid*

(Fig. 60)
ECHINOIDS are still in existence today as the free-moving 'Sea-Urchins'. The body comprises a skeletal structure, under a thin tissue layer of thin, brittle, limy plates in vertical rows called the 'test', the size of each plate being increased as the marine animal grows. These tightly fitting and interlocking plates in five double rows, the interambulacral areas, alternate with five double rows of smaller plates, the ambulacral areas. The interambulacral plates have sockets for the numerous moveable, protective spines; the interambulacral plates and ambulacral plates also bear smaller spines. The ambulacral plates have rows of pores through which the soft tube-feet are projected, either for movement on the oral side or breathing on the aboral side.

The mouth, surrounded by the periostome, is sited on the oral or underside of the body and has five 'teeth' in a complex structure called 'Aristotle's Lantern', except for the Heart Urchins and Sand Dollars which do not have this structure. On the test's upper or aboral side, at the apex, is the apical disc. This has in its centre the periproct in the centre of which is the anus or vent. Sea-Urchins are divided into two sub-classes–the *Regular*, with a test having a five-fold radial symmetry, and the *Irregular*, with a test having a varying degree of bilateral symmetry. The tests are globular, as in the Common Sea-Urchin, flattened as in the Sand Dollar or Cake Urchin, or heart-shaped as in the Heart Urchins. Dead examples, with all the spines missing, are cast ashore and these, with their beautiful patterning, make it easier to understand their construction, the rows of wart-like sockets indicating where the spines were formerly attached. Shell shops frequently sell them quite cheaply.

Some of the fossil Sea-Urchins, which occur from the Ordovician Period to Recent Period, are common; others are less often found. For example, fossil *Echinoids* are abundant in the limestones and chalks but usually rare in clays. The fossil collector searching for *Echinoid* fossils should remember that the examples are usually not found spined as the spines break off after death and so will comprise the shape of the 'shell' or test. Exceptions to this may be *Cidaris, Hemicidaris*, which may be found with a few tough spines attached to the tubercles. Even so, all *Echinoids* are easily identified and cannot be confused with other fossils when entire. It should also be remembered that the test may have disintegrated into fragments prior to fossilisation and may be flattened or scattered in the enclosing material. The numerous broken spines are also found fossilised. In Chapter Two, I refer to how Sea Urchins became fossils.

CIDARIS (Fig. 61)
Triassic to Recent. A Regular *Echinoid*. Test spherical, with a slightly flattened apex; ambulacral areas long, slender, with a mosaic of polygonal plates, large tubercles and spines in between; large mouth at centre of test. Up to 3 inches in diameter. The related *Hemicidaris*, Jurassic, is similar, but has a hemispherical shape due to its flattened lower or dorsal surface.

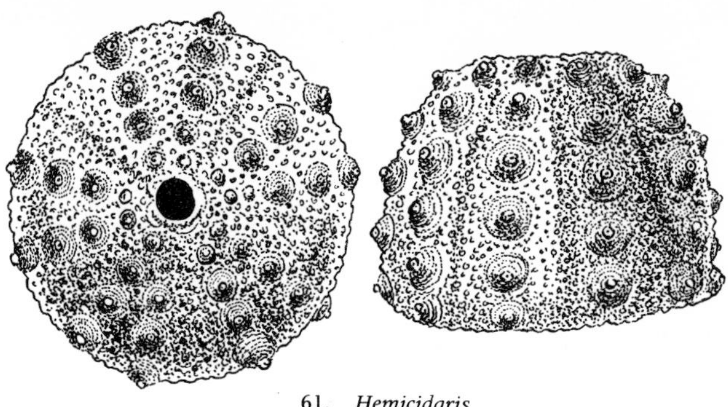

61. *Hemicidaris*

TYLOCIDARIS

Another *Cidarid*, Cretaceous, which is similar, with a mosaic of polygonal plates and large tubercles, but the large spines are club-like in shape.

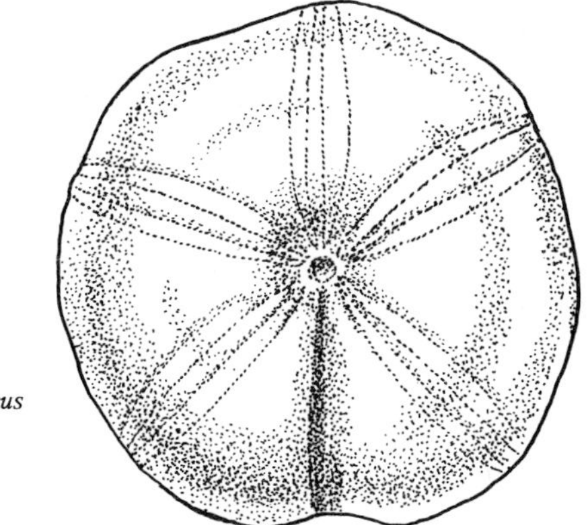

62. *Clypeus*

CLYPEUS (Fig. 62)

Jurassic-Inferior Oolite. An Irregular Sand Dollar. Test flattened, rounded or five-sided, has a surface of very tiny tubercles; ambulacral areas like five flower petals; mouth at centre of test. Up to 4 inches in diameter.

CONULUS (Fig. 63)

Cretaceous-Chalk. An Irregular Echinoid. Test conical, rounded, has a smooth surface; ambulacral areas like five flower petals; mouth at centre of test. Up to $1\frac{1}{2}$ inches in diameter.

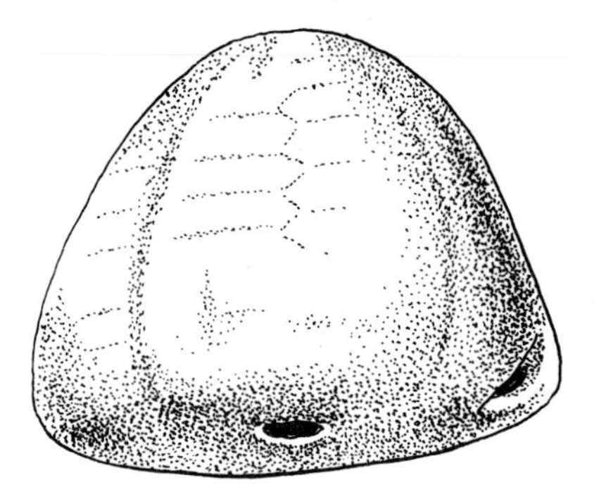

63. *Conulus*

ECHINOCORYS (Fig. 64)

Upper Cretaceous-Chalk. A Regular Echinoid. Test globular, with a flattened apex; oval; ambulacral areas are weak, like five flower petals; mouth near anterior margin. Up to 3 inches in length.

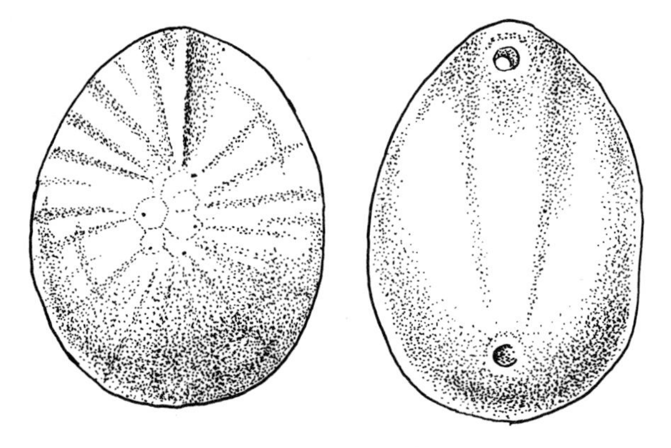

64. *Echinocorys*

LOVENECHINUS (Fig. 65)

Mississippian (North America), Carboniferous (British Isles). A Regular Echinoid. Test spherical; ambulacral areas long, narrow, pointed, and each comprise four columns of small plates; in between these are up to seven columns of large, six-sided plates; mouth central. Up to 4 inches in diameter.

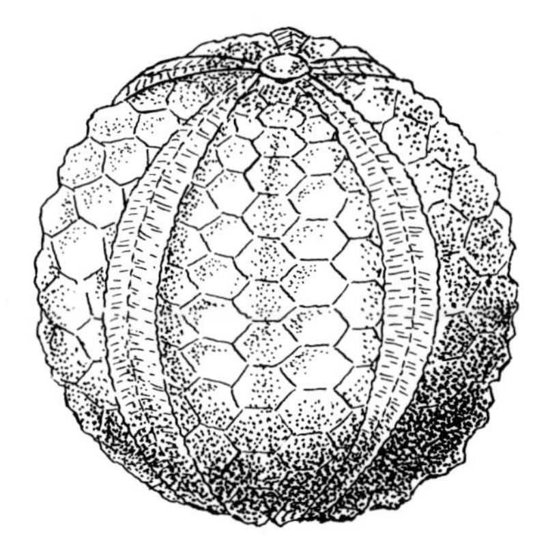

65. *Lovenechinus*

MICRASTER (Fig. 66)
Cretaceous to Miocene. An Irregular Echinoid. Test heart-shaped, has a granular surface; ambulacral areas sunken, forming five flower-petal shape; in between these are narrow plates; mouth near anterior margin. Up to 2 inches in length.

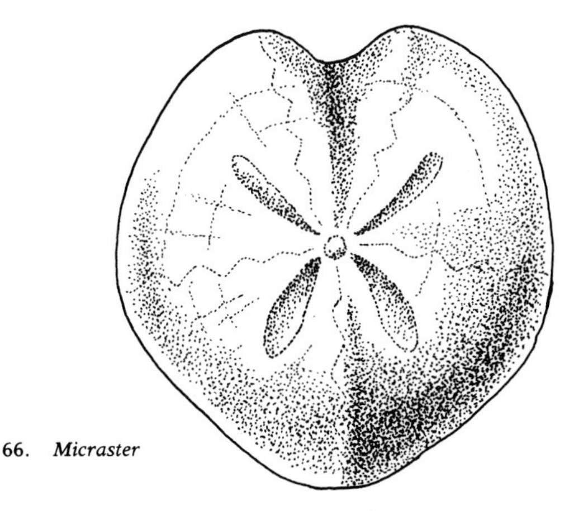

66. *Micraster*

HOLOTHUROIDS still exist today as the Sea Cucumbers, with a long, leathery gherkin-like body. The mouth is situated at one end surrounded by tentacles for feeding. Although there is no strong internal skeleton there are small limy plates or spicules of various shapes in the skin. These spicules occur as micro-fossils and are identified from other micro-fossils by their distinctive dog-bone, cross, circular, wheel, etc., shape. *Holothuroids* occur as fossils from the Cambrian Period.

MOLLUSCA

The *Mollusca* is a very large group of soft-bodied *Invertebrates* that vary considerably in shape and size, from the symmetrical Bivalves, coiled Univalves, to the plated Chitons and internally shelled Cephalopods. For space reasons and because this book is concerned with fossils it is not intended to dwell closely on their physical differences, but these are entirely described in my book 'Coast, Estuary & Seashore Life' (Gifford). Some of these species which are extremely common today, particularly the marine species, also occur as fossils, but other living species, which are also equally numerous now in the wild, are rare as fossils. The fossil *Mollusca* are divided into five classes–*Amphineura* (Placophora); *Scaphopoda, Cephalopoda, Gastropoda, Pelecypoda* or *Lamellibranchia*. The *Amphineura* (Placophora) which occur today as the flattened, oval-shaped, flexible, eight-shell-plated marine Chitons, are rare as fossils, from the Lower Palaeozoic Era to the Recent Period. They are very small and relatively unimportant and are unlikely to be found, so I have not gone into any detail about them.

Scaphopoda

The *Scaphopoda* are still in existence today as the marine Tusk Shell Molluscs which have a wider tubular shell resembling an elephant's tusk, being wider at the anterior end from which the Mollusc's body projects as it partially buries itself in the mud and sand. The most frequent species occurring in British and North American waters is the slightly curved ivory-white *Dentalium entale*.

DENTALIUM ANTALIS
Ordovician to Recent Period, particularly Upper Eocene Period. Shell typically tusk-shaped, tapered, curved. Up to 5 inches in length.

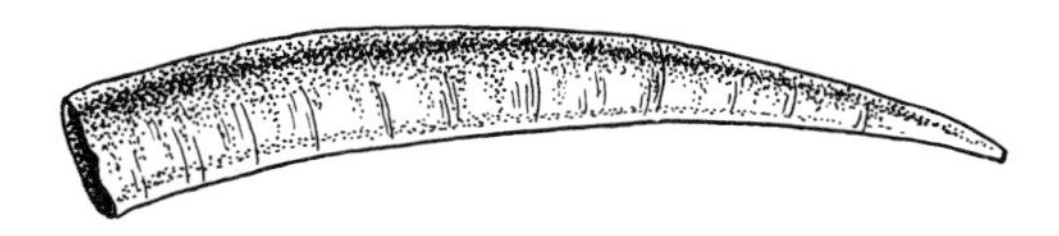

67. *Dentalium*

DENTALIUM DECUSSATUM (Fig. 67)
Cretaceous-Gault. Shell typically tusk-shaped, tapered, with a pronounced curve to narrow end. Up to 3 inches in length.

Cephalopoda

The *Cephalopoda*, 'head-footed', exist today as the familiar *Octopoda* (Octopus), *Decapoda* (Squids and Cuttlefish) and *Tetrabranchiata* (Nautilus) and live entirely in the sea. The fossil *Cephalopoda* are divided into two groups–the *Tetrabranchiata* and *Dibranchiata*. These are sub-divided into: *Tetrabranchiata-Nautiloidea, Ammonoidea;* and *Dibranchiata-Belemnoidea, Sepioida, Teuthoida, Octopoida.*

TETRABRANCHIATA
NAUTILOIDEA
The only living example is the Pearly Nautilus which occurs in the south-west Pacific Ocean, its smooth, coiled shell being divided into numerous chambers, up to 36, by 'septa' or transverse plates, through which extends a 'siphuncle' or fleshy tube. Where the 'septa' meets the interior of the shell outer wall is formed the curving line called the 'suture'. Fossil *Nautiloids* occur from the Cambrian to Recent Period and some examples are widespread and numerous. The shells are either straight, curving or coiled, usually smooth, flat or bulbous, some of the early examples being very large.

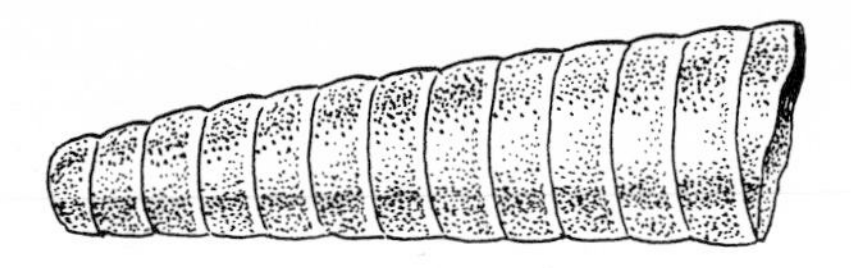

68. *Bactrites*

BACTRITES (Fig. 68)
Ordovician to Permian. Shell narrow, straight, with a rounded cross-section. The suture lines are simple, weak; septa edges folded and the siphuncle near the outer or ventral margin; has a small ventral lobe. Up to $1\frac{1}{2}$ inches in length.

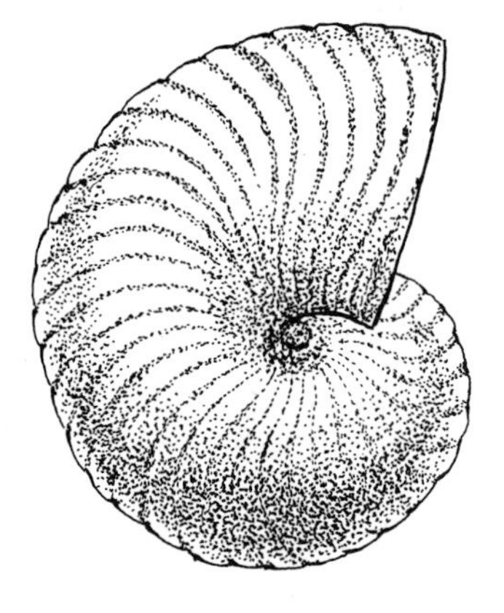

69. *Cymatoceras*

CYMATOCERAS (Fig. 69)
Cretaceous-Lower Greensand. Shell tightly coiled; similar to *Nautilus*, outer whorls being very much larger than inner whorls, but has many more curving growth lines that are also more prominent; large aperture. Up to $1\frac{1}{2}$ inches in diameter.

CYRTOCERAS
Ordovician to Devonian. Shell, curved, tusk-like, with a rounded cross-section; siphuncle ventral and prominent. Up to 3 inches in length.

DAWSONOCERAS (Fig. 70)
Middle Silurian to Lower Devonian. Shell straight, with a

Plate VII Pipe Rock. Base of the Cambrian, Sutherland, Scotland.

Plate VIII *Dactylotheca plumosa* (Artis, 1825) and *Asterotheca oreopteridia* (Schlotheim, 1804). Ludlow's Pit, Radstock, Somerset, England. Upper Coal Measures–Carboniferous. 'Pecopteris' is a form Genus.

rounded cross-section; siphuncle central and small; growth rings resemble the thread of a screw. Up to 5 inches in length.

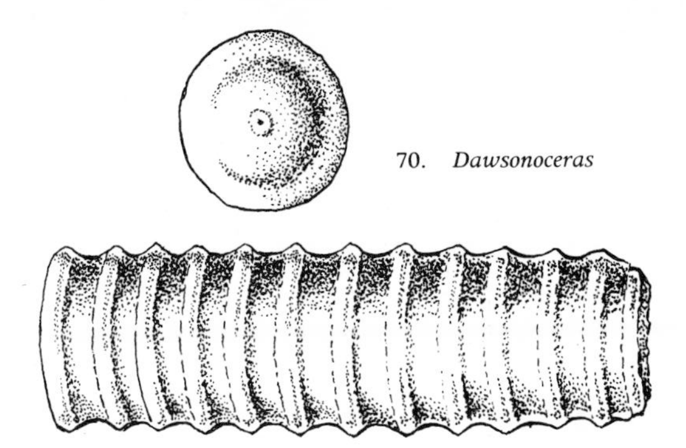

70. *Dawsonoceras*

DOLORTHOCERAS

Devonian to Pennsylvanian (North America), Carboniferous (British Isles). Shell straight, with a rounded-oval cross-section; the sutures are transverse; siphuncle is central. Up to 4 inches in length.

ENDOCERAS

Ordovician. Shell straight, long, conical, with a rounded-oval cross-section. The suture lines are straight; shell walls thick; siphuncle large with a layer of thick calcareous material along its length. Very large examples occurred, up to 13 feet in length.

71. *Gomphoceras*

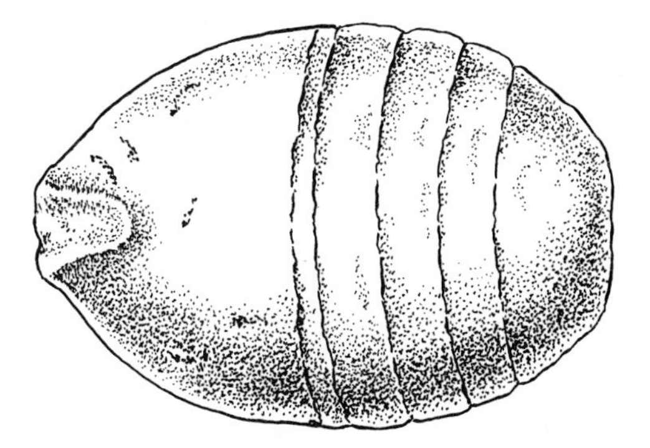

GOMPHOCERAS (Fig. 71)
Ordovician to Devonian. Shell oval, bulbous, rounded in cross-section; body chamber large, smooth; sutures weak; septa simple. Up to 3 inches in length.

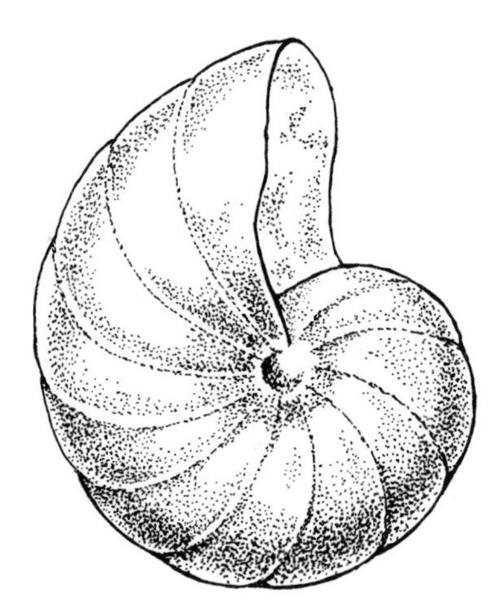

72. *Nautilus*

NAUTILUS [GENOCERAS] (Fig. 72)
Middle Jurassic-Inferior Oolite. Very similar to present-day Pearly Nautilus; Shell tightly coiled, smooth; outer whorls are much larger than inner whorls; curving growth lines weak; very large aperture. Up to $2\frac{1}{2}$ inches in diameter.

ORTHOCERAS
Lower Carboniferous. Shell straight, long conical, with a rounded cross-section; suture lines straight; siphuncle central. Up to 3 inches in length.

PHRAGMOCERAS (Fig. 73)
Silurian. Shell curved, laterally compressed, with a rounded-oval cross-section; siphuncle on concave side; transverse growth lines; aperture shaped like a figure 8; body chamber large. Up to 5 inches in length.

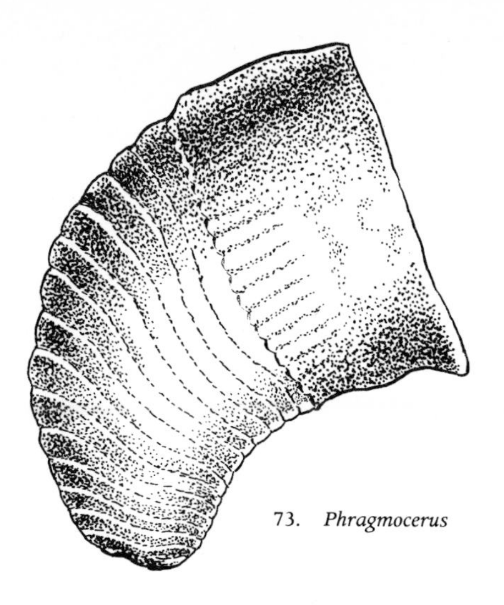

73. *Phragmocerus*

AMMONOIDEA

The *Ammonoidea*, which contain some of the most familiar fossils, are now all extinct. They developed in the Devonian Period, from a *Nautiloid*, probably *Bactrites*, and died out by the end of the Cretaceous Period. They are divided in order of development into the *Goniatites*, with sharply angular suture lines, from Devonian to the Permian Period; *Ceratitoids*, with suture lines having smooth rounded saddles and finely denticulated or frilled lobes, Devonian to Triassic; and from them arose the *Ammonites*, with very variable suture lines having frilled saddles and lobes, Jurassic and Cretaceous Periods. The *Ammonites* and *Ceratitoids* often occur in abundance in their Period rocks in the British Isles, but the *Goniatites* less frequently occur and may be impressions of the crushed shell.

GONIATITES

AGONIATITES

Middle Devonian. Shell tightly coiled, laterally flattened, growth lines weak; siphuncle ventral; suture simple. Up to 6 inches in diameter.

GASTRIOCERAS (Fig. 74)
Upper Carboniferous-Coal Measures (British Isles), Pennsylvanian (North America). Shell tightly coiled, laterally flattened; growth lines weak; suture has simple folding; prominent ribs near the dorsal or inner margins of the whorls having short tubercles. Up to $1\frac{1}{2}$ inches in diameter.

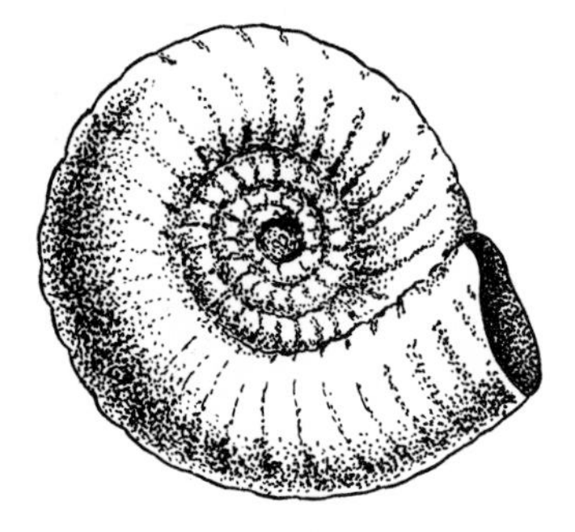

74. *Gastrioceras*

GONIATITES (Fig. 75)
Lower Carboniferous (British Isles), Mississippian (North America). Shell tightly coiled, globular, smooth; deeply wavy suture lines, similar to that of *Muensteroceras*, but deep ventral lobe has angled sides. Up to 1 inch in diameter.

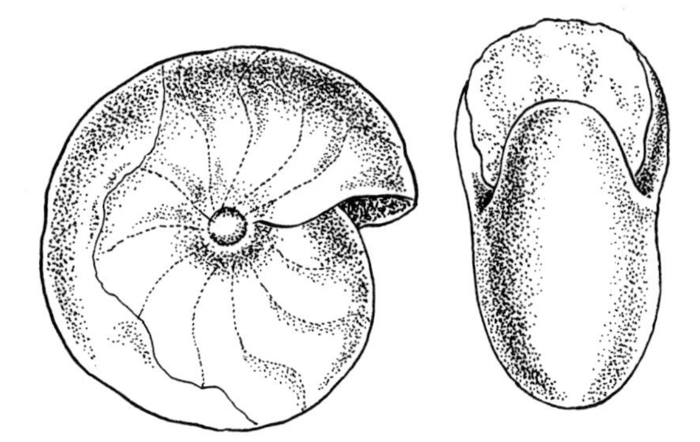

75. *Goniatites*

MUENSTEROCERAS

Carboniferous (British Isles), Mississippian (North America). Shell tightly coiled, laterally flattened or slightly globular; smooth; wavy suture lines have deep ventral lobe, with straight sides (see *Goniatites*). Up to 1 inch in diameter.

RETICULOCERAS

Carboniferous-Millstone Grit. Shell tightly coiled, laterally flattened or slightly globular; identified by the longitudinal and transverse lines forming a finely striated ornamentation on the outer whorl. Up to 1 inch in diameter.

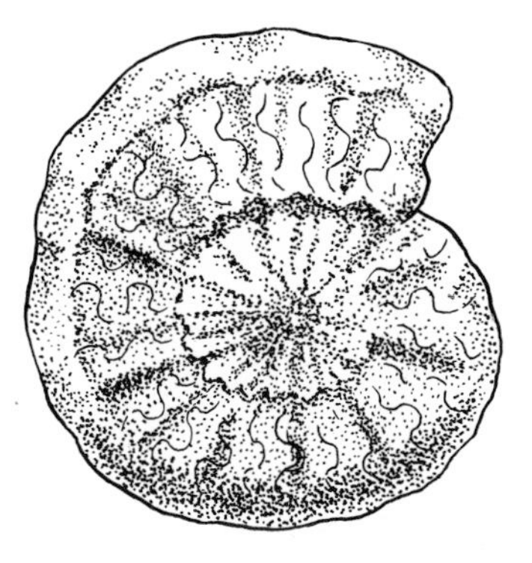

76. *Ceratites*

CERATITOIDS

CERATITES (Fig. 76)

Middle Triassic. Shell tightly coiled, laterally flattened; ribs weak and do not reach the flat or raised margin of the outer whorl; wavy suture lines blunt lobed, quickly identify it; last, outer whorl weakly encloses small whorls. Up to 2 inches in diameter.

MEEKOCERAS

Lower Triassic. Shell tightly coiled, laterally flattened, smooth; outer whorl has a flat margin; wavy suture lines have secondary folding in the blunt lobes. Up to 2 inches in diameter.

SAGENITES (Fig. 77)

Upper Triassic. Shell tightly coiled, globular but slightly laterally flattened; whorls have tubercle-like nodes forming spiral and radial ornamentation; some short spines may have survived; suture lines very irregular. Up to 3 inches in diameter.

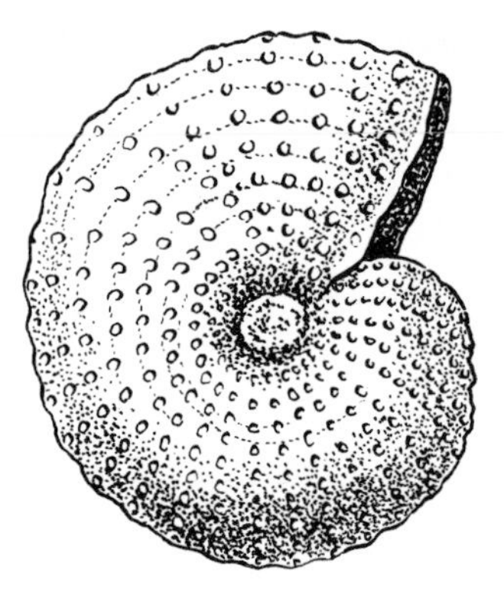

77. *Sagenites*

AMMONITES

ACANTHOSCAPHITES

Upper Cretaceous. Shell tightly coiled, except for outer whorl which is free near venter and expanded away from coil; ribs prominent with several small and large nodes; suture lines very complex and irregular. Up to 4 inches in diameter.

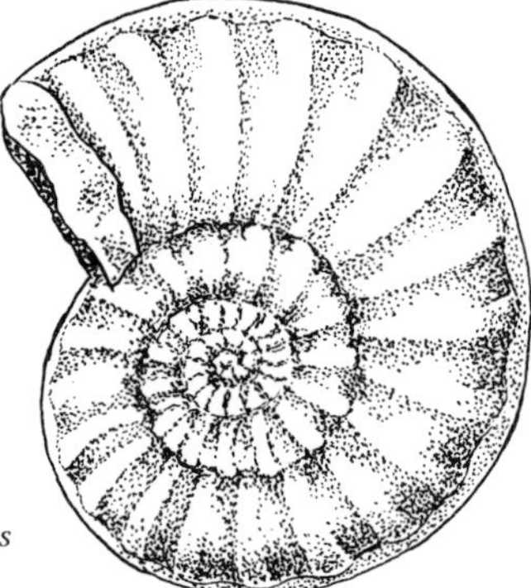

78. *Asteroceras*

ASTEROCERAS (Fig. 78)

Lower Jurassic-Lower Lias. Shell tightly coiled, laterally

flattened; numerous ribs prominent; outer whorl has a groove each side of the raised keel. Up to 2 inches in diameter.

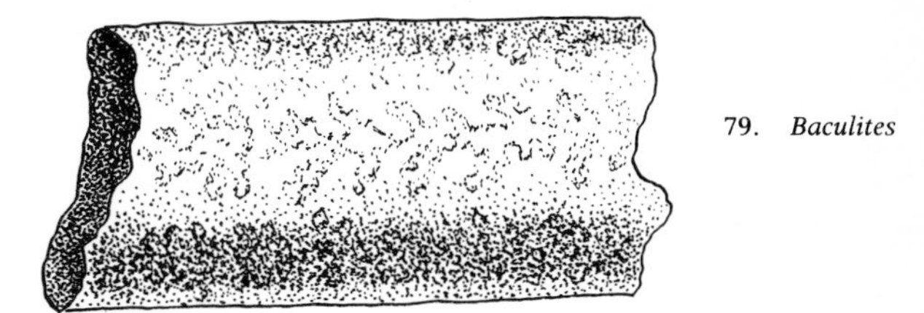

79. *Baculites*

BACULITES (Fig. 79)
Upper Cretaceous. Shell straight, tapering, except for initial stage of shell which has a short spiral curve; smooth, or has weak, rounded ribs, or curving striae; suture lines symmetrical with complex folding, fern or leaf-like; only mid-section may be found so must be identified from suture lines. Extremely long examples, up to 6 feet, occurred, but up to 6 inches in length more frequent.

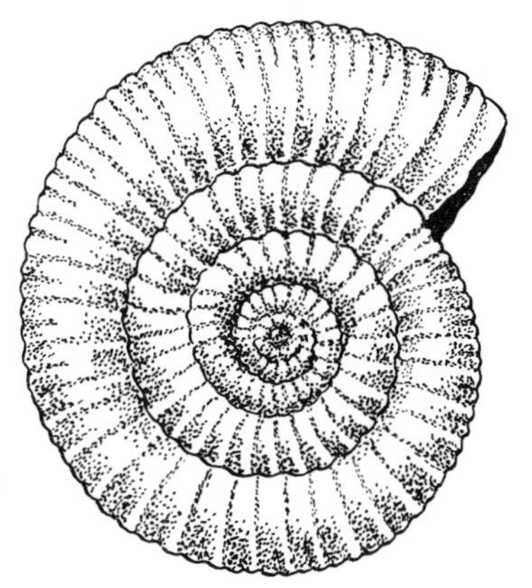

80. *Dactylioceras*

DACTYLIOCERAS (Fig. 80)
Jurassic-Upper Lias. Shell tightly coiled, laterally flattened; numerous straight ribs; branched ribs on rounded outer whorl; suture complex. Up to 3 inches in diameter.

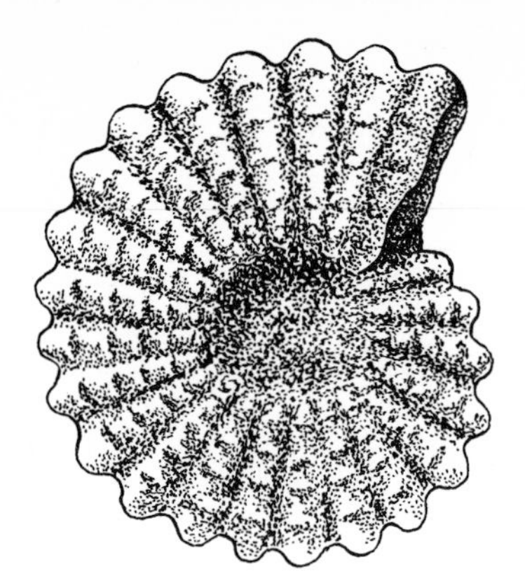

81. *Douvilleiceras*

DOUVILLEICERAS (Fig. 81)
Cretaceous-Lower Greensand, Gault. Shell tightly coiled, laterally flattened; central whorls forming a deep depression surrounded by a large, rounded outer whorl; ribs straight, large and prominent, with numerous tubercles; may also be spined. Up to 2 inches in diameter.

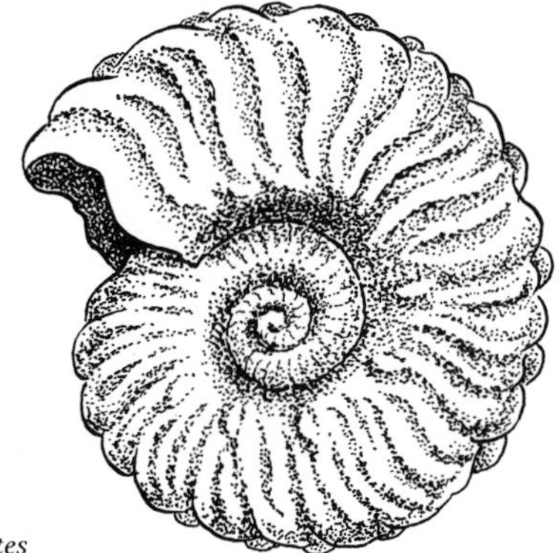

82. *Euhoplites*

EUHOPLITES (Fig. 82)
Lower Cretaceous-Gault. Shell tightly coiled, laterally flattened; inner central whorls form a shallow depression; prominent ribs curved to large nodes on the outer margin of the large, outer whorl; outer whorl having a shallow groove in between nodes. Up to 2 inches in diameter.

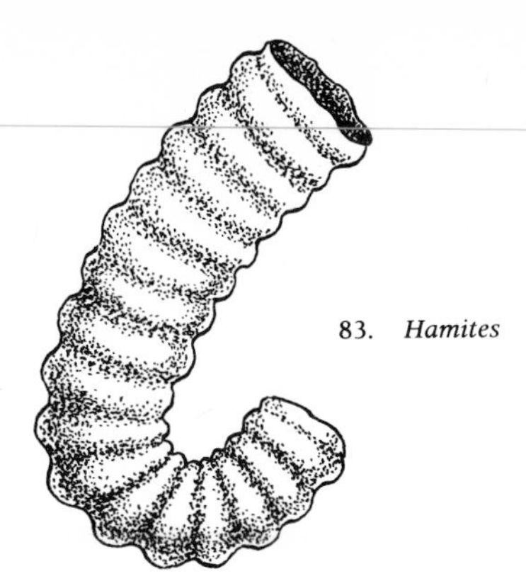

83. *Hamites*

HAMITES (Fig. 83)
Lower Cretaceous-Gault. Shell in varying degrees of uncoiling; fossil shell when found may have a long shaft with a curved shorter shaft at one end only or have a curving shaft at each end, one end being tapered; rounded cross-section; ribs prominent; suture lines irregular and complex. Up to 3 inches in length.

HARPOCERAS
Jurassic-Upper Lias. Shell tightly coiled, laterally flattened; smooth; numerous ribs sickle-shaped; outer whorl has prominent raised keel. Up to 3 inches in diameter.

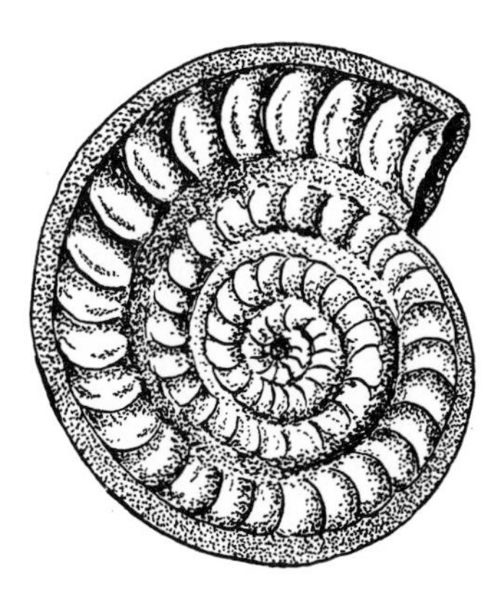

84. *Hildoceras*

HILDOCERAS (Fig. 84)
Jurassic-Upper Lias. Shell tightly coiled, laterally flattened; inner whorls form a wide, shallow depression; ribs short, curved; outer whorls ventral margin has a slight groove each side of the prominent raised keel and a groove on its inner margin so that it is quadrangular in cross-section; suture lines irregular. Up to 3 inches in diameter.

OXYNOTICERAS
Jurassic-Lower Lias. Shell tightly coiled, laterally flattened; outer whorl's ventral margin angular, high and sharp edged, with the inner or dorsal margin overlapping and almost covering the inner whorls; ribs straight on inner margin but curve towards outer whorl's ventral margin. Up to 2 inches in diameter.

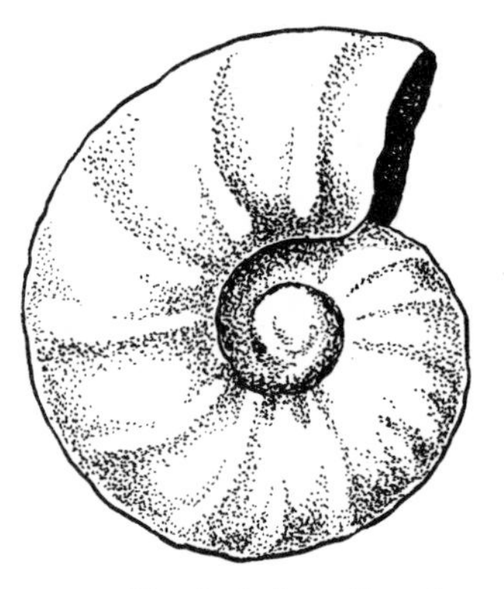

85. *Pachydiscus (Lewesiceras)*

PACHYDISCUS (Fig. 85)
Cretaceous-Upper Chalk. Shell tightly coiled, laterally flattened; smooth; ribs weak, slightly curved; outer whorl's ventral margin rounded. The largest *Ammonite* recorded; up to $6\frac{1}{2}$ feet in diameter.

PARKINSONIA (Fig. 86)
Jurassic-Inferior Oolite. Shell tightly coiled, laterally flattened; numerous ribs straight or slightly curved on inner and outer whorls, but branched on outer whorl's ventral margin which is also grooved. Up to 4 inches in diameter.

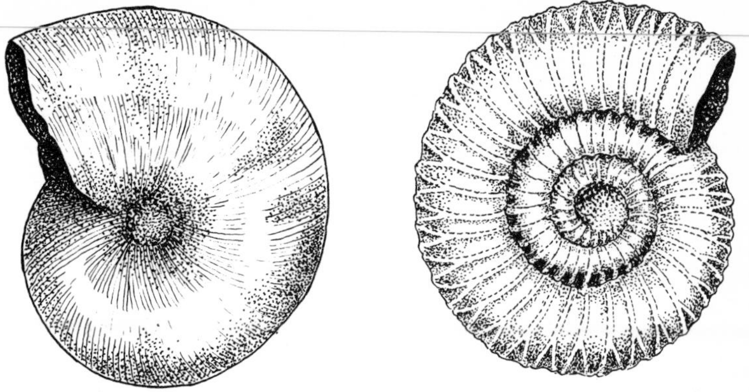

87. *Phylloceras*

86. *Parkinsonia*

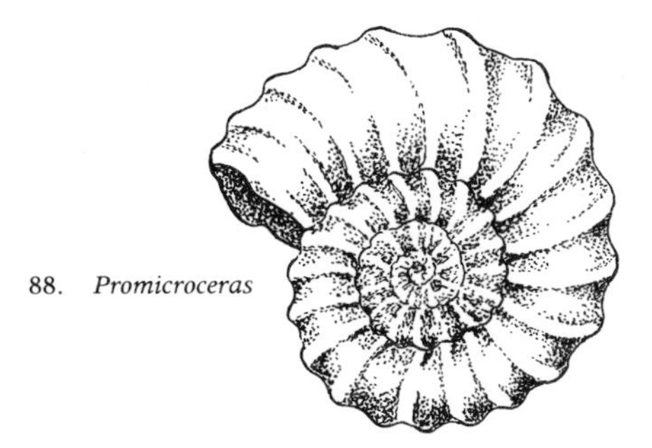

88. *Promicroceras*

89. *Scaphites*

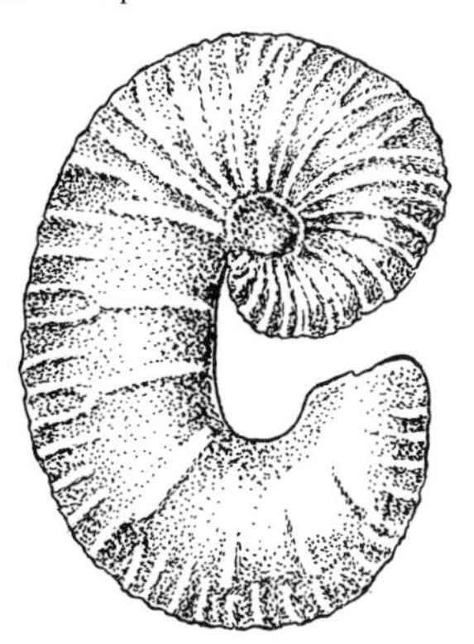

90. *Stephanoceras*

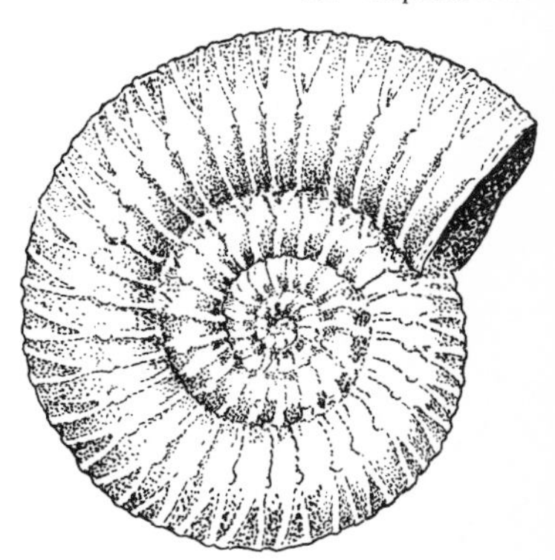

PHYLLOCERAS (Fig. 87)
Lower Jurassic-Upper Lias. Shell tightly coiled, laterally flattened; outer whorl very large, the inner margin overlapping and almost covering inner whorls; ribs numerous but very fine; suture lines irregular and very complex, fern or leaf-like. Up to 5 inches in diameter.

PROMICROCERAS (Fig. 88)
Lower Jurassic-Lower Lias. Shell tightly coiled, laterally flattened; prominent ribs straight or slightly curved, curving over ventral margin of body whorl. Up to 2 inches in diameter.

SCAPHITES (Fig. 89)
Cretaceous-Lower Chalk. Shell in varying degrees of uncoiling; usually flattened; first spiral whorls linked to short curving shaft, but last, outer whorl may be free at hooked end; has fine and more prominent ribs some of which may be branched, several being tubercled; suture lines irregular and complex. Up to 2 inches in length.

STEPHANOCERAS (Fig. 90)
Jurassic-Inferior Oolite. Shell tightly coiled, laterally flattened; outer whorl slightly embraces inner whorl near venter; ribs straight but slightly curved over ventral margin of outer whorl; prominent nodes in central position of outer and inner whorls. Up to 5 inches in diameter.

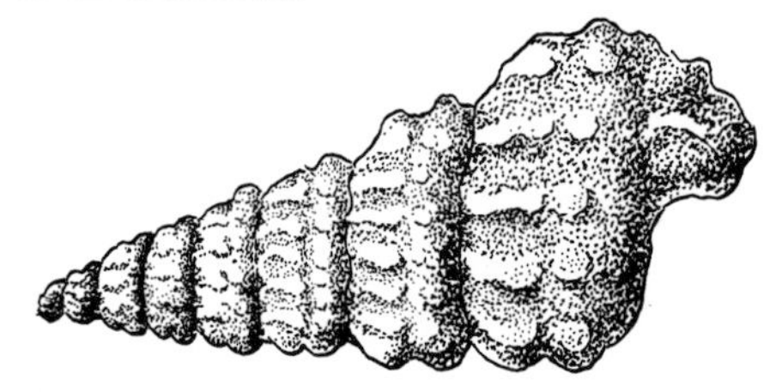

91. *Turrilites*

TURRILITES (Fig. 91)
Cretaceous. Shell coil helicoid or coiled like a *Gastropod*, tapering to a spire; prominent ribs transverse, some being

tuberculed; suture lines very irregular and complex. Up to 5 inches in length.

DIBRANCHIATA

BELEMNOIDEA

The *Belemnoidea* were carnivorous marine creatures similar to the present-day Cuttlefish, with a head bearing suckered tentacles and a soft body that contained a crystalline lime skeleton or guard. These guards are common as fossils and were, in the past, thought by people who found them to be 'the Devil's thunderbolts', the name *Belemnites* meaning 'a dart'. They are cigar or bullet-shaped and in cross-section have radiating calcareous layers similar to the 'rings' of a tree trunk. One end, the posterior, is sharp- or blunt-pointed, while the other, anterior, end has a depression, the alveolus, which may be filled by a conical structure or chambered shell, the phragmocone, to which, on one side, is sited a projection called the pro-ostracum. The latter usually exists only on guards that are extremely well-preserved. *Belemnites* vary, depending on the Period, and may have the phragmocone projecting from or enclosed within the guard. The shape may be the same diameter along its length except for a conical posterior end, or have a bulbous tip at the posterior end. The guards may also have a furrow or longitudinal grooves, more defined on the anterior end. The cross-section shape is also important in identification. *Belemnites* occur as fossils from the Carboniferous (British Isles), Mississippian (North America), to the end of the Cretaceous Period finally becoming extinct in the Eocene Period. They are particularly prolific in the Jurassic clays and Cretaceous chalk.

BELEMNITELLA (Fig. 92)

Upper Cretaceous-Upper Chalk. Circular cross-section, widest at anterior end, posterior end having small bulbous tip; has long grooves on one side; also branched patterning on surface. Up to 3 inches in length.

HIBOLITES (Fig. 93)

Cretaceous-Speeton Clay. Circular cross-section. Varies in shape, may be sharp-pointed, bulbous at posterior end, tapering to anterior end, or sharp-pointed but same width until slight taper at anterior end. Up to 3 inches in length.

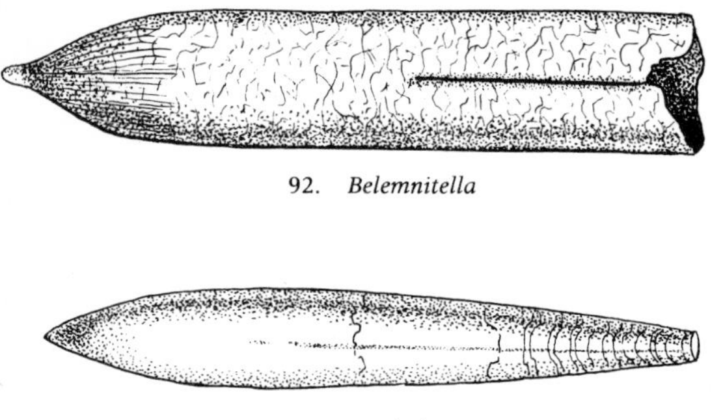

92. *Belemnitella*

93. *Hibolites*

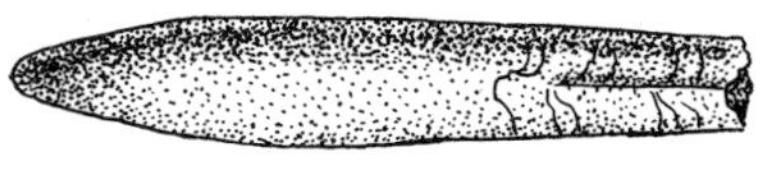

94. *Neohibolites*

NEOHIBOLITES (Fig. 94)
Cretaceous-Gault. Circular cross-section. Cigar-shaped, with sharp or blunt-pointed posterior end; grooved. Up to 2 inches in length.

PACHYTEUTHIS
Jurassic, Cretaceous. Sub-oval cross-section widest at anterior end, tapering to sharp-pointed posterior end; grooved on one side. Up to 4 inches in length.

PSEUDOHASTITES (Fig. 95)
Jurassic-Upper Lias. Circular cross-section; very thin, slender; same diameter along length, except for tapering, conical posterior end and wider anterior end; grooved. Up to 5 inches in length.

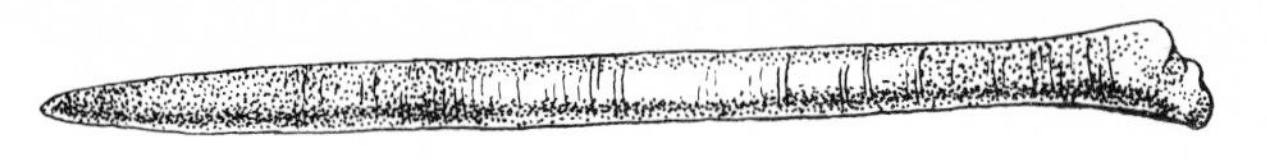

95. *Pseudohastites*

SEPIOIDA

The present-day *Sepia* or Cuttlefish may have developed from the *Belemnoids*, as varying development stages have been discovered revealing the change-over from the internal guard and conical phragmocone to a larger, flatter phragmocone and finally the *Sepia's* 'shell' or cuttlebone. Fossil *Sepioids* are scarce. One example, *Belosaepia sepioidea*, has been found in the Bracklesham Beds and London Clay Series of the Eocene Period in the British Isles.

TEUTHOIDA, OCTOPOIDA

The *Teuthoids* or present-day true Squids may have also developed from the *Belemnoids*. They have a long, slender, horny 'shell' or 'pen', so-called because they resemble a quill pen, as an internal skeleton and these have been discovered in Jurassic and Cretaceous rocks, but are rare. The *Octopoida* also probably developed from the *Belemnoids* in the Mesozoic Era, but very little is known about this at the present time, as their fossil remains are very rare.

Gastropoda

The *Gastropoda* exist today as a very large group of marine and land, limy-shelled Univalve Molluscs. It is not possible to describe here all the various living forms and their anatomy, but these are entirely covered in Chapter 2 of my companion volume 'Collecting World Sea Shells' (Bartholomew). Fossil *Gastropods* occur from the Cambrian Period to Recent Period. Many of them, such as *Aporrhais, Crepidula, Littorina, Conus, Natica, Fusus, Turritella,* etc., have present-day descendants.

APORRHAIS (Fig. 96)

Pleistocene-Red Crag. Shell spired, with several rounded whorls having traces of the numerous tubercles; elongated aperture has an expanded, curving, outer lip. Up to $2\frac{1}{2}$ inches in length.

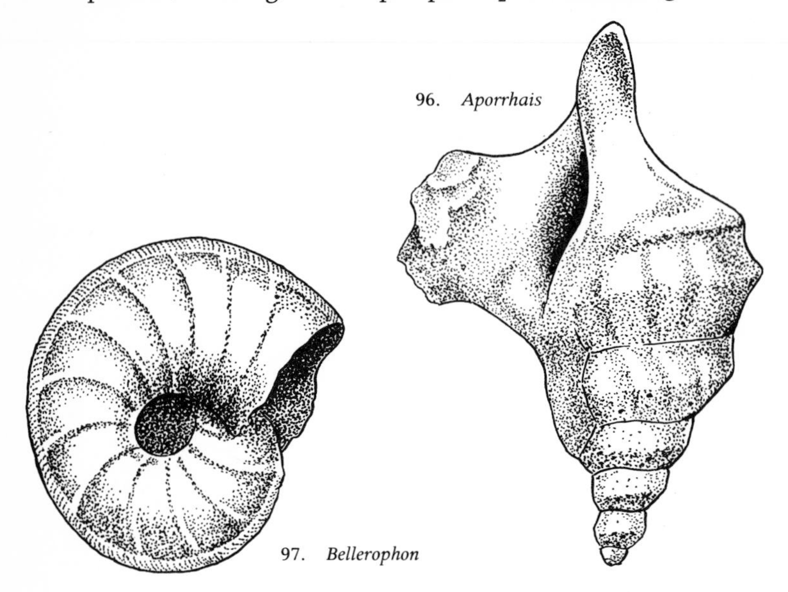

96. *Aporrhais*

97. *Bellerophon*

BELLEROPHON (Fig. 97)
Ordovician to Triassic. Shell coiled, globular; outer whorl partly overlapping inner whorls; weak ribs; large, wide aperture; prominent slit band from aperture around outer whorl. Up to 2 inches in diameter.

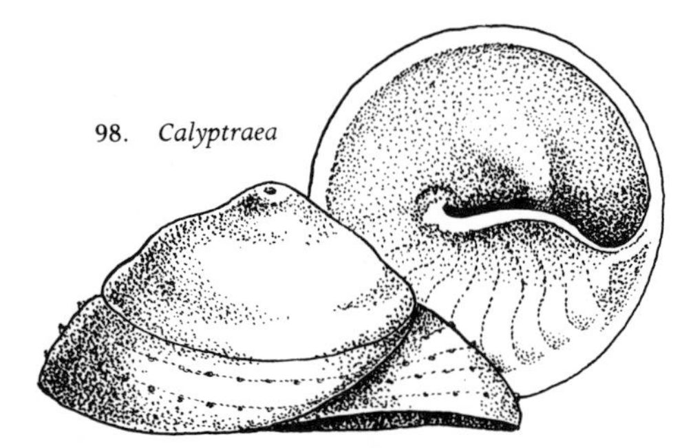

98. *Calyptraea*

CALYPTRAEA (Fig. 98)
Eocene-Barton Beds, to Recent. Shell round, conical, but may be almost flat, sometimes with a central nipple-like apex; spiral growth lines; interior has a ledge or plate on posterior half. Up to 1 inch in diameter.

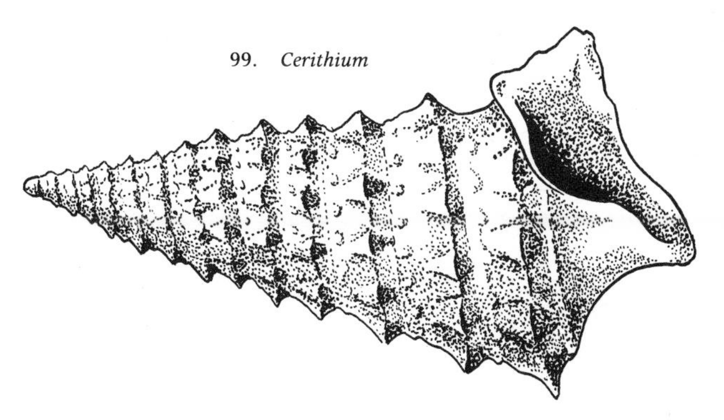
99. *Cerithium*

CERITHIUM (Fig. 99)
Jurassic to Recent. Shell tall, pointed spired, with numerous noduled whorls; aperture rounded with siphon canal. Up to 5 inches in length.

CONUS
Cretaceous to Recent. Shell has a short, sometimes flattened, but usually conical, spire; very large, tapering, body whorl has a long, narrow aperture. Up to 2 inches in length.

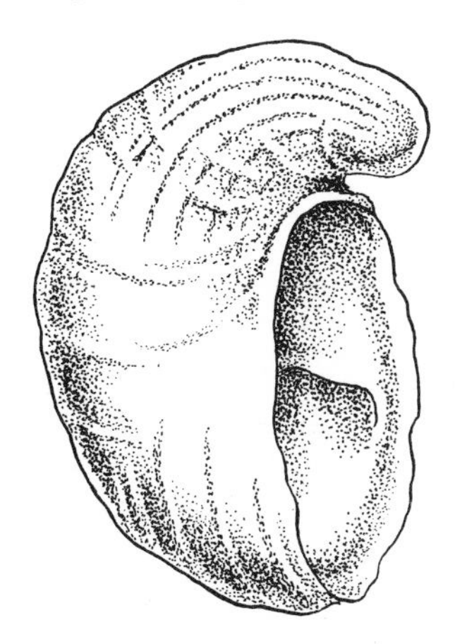
100. *Crepidula*

CREPIDULA (Fig. 100)
Cretaceous to Recent. Shell oval, but shape can vary, flat, taller or curving; beak twisted; aperture large, with an interior horizontal ledge or platform. Up to 2 inches in length.

EMARGINULA
Pleistocene-Red Crag. Shell Limpet-shape, ribbed and cross-lined from apex to margin; apex distinctly curved; identified also by vertical slit or marginal notch on anterior edge, originally part of the respiratory system. Up to ½ inch in length.

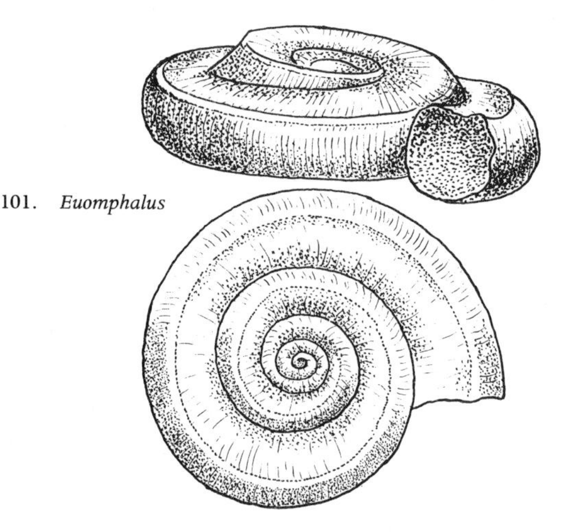
101. *Euomphalus*

EUOMPHALUS [STRAPAROLLUS] (Fig. 101)
Carboniferous-Limestone. Shell tightly coiled, laterally flattened; whorls rounded or triangular. Up to 2 inches in diameter.

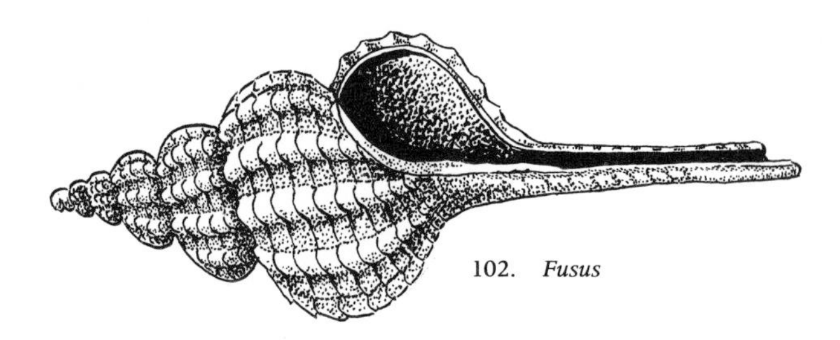
102. *Fusus*

FUSUS (Fig. 102)
Cretaceous to Recent. Shell elongated, slender, with a tall spire and rounded, ridged whorls; aperture rounded with a very long siphon canal. Up to $2\frac{1}{2}$ inches in length.

HELCIONELLA
Lower Cambrian. Shell Limpet-shape; flat-coned, with apex close to anterior margin. Up to 1 inch in length.

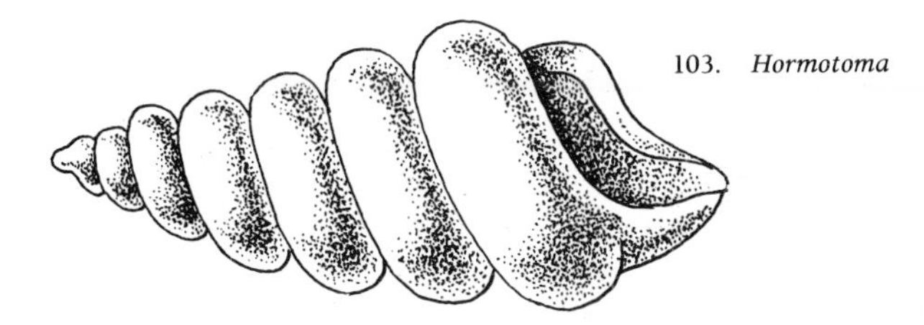
103. *Hormotoma*

HORMOTOMA (Fig. 103)
Ordovician to Silurian. Shell elongated, slender, with a tall spire and rounded, smooth whorls with a deep suture; aperture notched. Up to $2\frac{1}{2}$ inches in length.

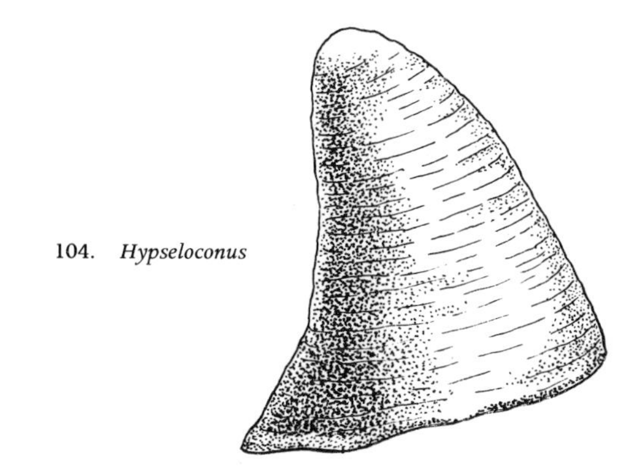
104. *Hypseloconus*

HYPSELOCONUS (Fig. 104)
Cambrian to Lower Ordovician. Shell Limpet-shape, conical; apex curved. Up to $\frac{1}{2}$ inch in height.

LITTORINA

Palaeocene to Recent. Shell Winkle-shaped, thick, rounded, up to 6 rounded whorls, the body whorl being the largest; spiral sculptured surface, but may be weak; aperture rounded. Up to 1 inch in height.

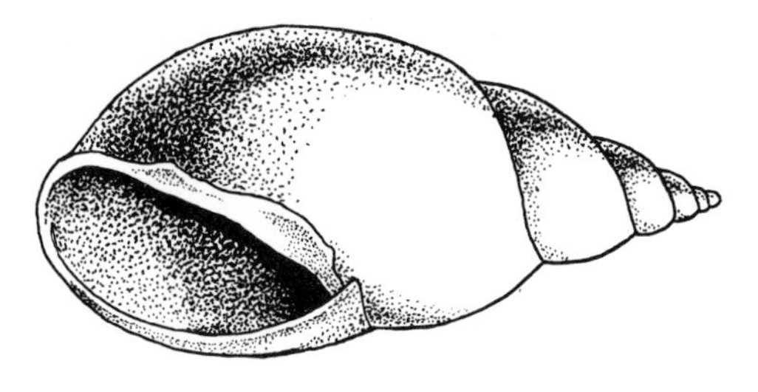

105. *Lymnaea*

LYMNAEA [GALBA] (Fig. 105)

Oligocene. Shell Whelk-shaped, but more slender; smooth; up to 6 rounded whorls, the body whorl being very large; aperture rounded-oval. Up to 1½ inches in length.

MACLURITES

Ordovician. Shell tightly coiled, laterally flattened; smooth; aperture oblong-oval. Up to 3 inches in diameter.

MURCHISONIA

Silurian to Permian. Shell elongated, with a tall spire; rounded or angular whorls, sometimes noduled; deep suture; aperture rounded-oval; outer lip has a sinus. Up to 2 inches in height.

MUREX [PTERYNOTUS] (Fig. 106)

Eocene-Barton Beds, to Recent. Shell blunt spired; up to 6 irregular whorls which may have blunt spines; large body whorl ridged; outer lip extended, bearing a spine or spines; aperture rounded with a siphon canal. Up to 2½ inches in length.

NATICA [AMPULLONATICA]

Eocene. Shell Snail-like, tightly coiled, bulbous; smooth; body whorl very large; oval aperture large. Up to 2 inches in height.

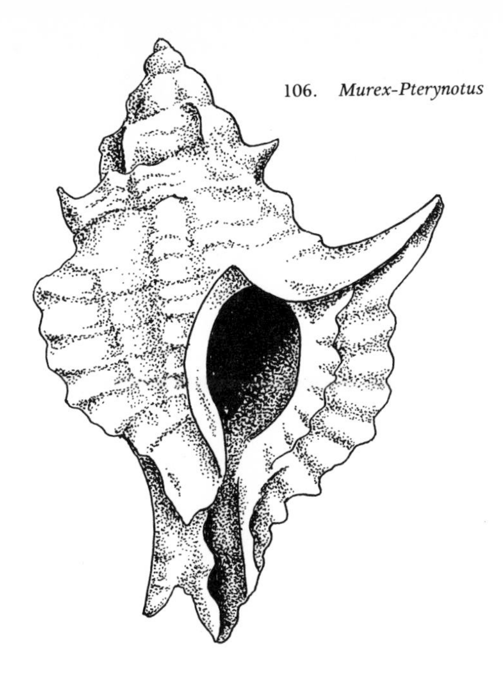

106. *Murex-Pterynotus*

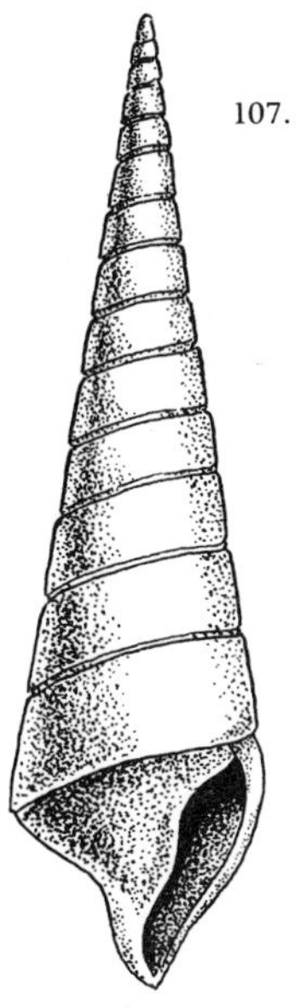

107. *Nerinea*

108. *Planorbis*

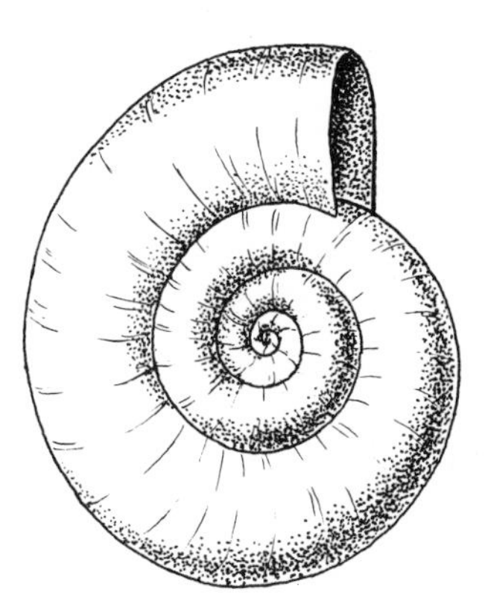

NERINEA [MELANIOPTYXIS] (Fig. 107)
Jurassic-Great Oolite, to Cretaceous. Shell elongated, slender, with a tall spire; whorls straight-sided; aperture oval with a short siphon canal. Up to 5 inches in height.

PLANORBIS (Fig. 108)
Jurassic to Recent. Shell tightly coiled, laterally compressed, discoidal; aperture rounded-oval. Up to $\frac{3}{4}$ inch in diameter.

109. *Platyceras*

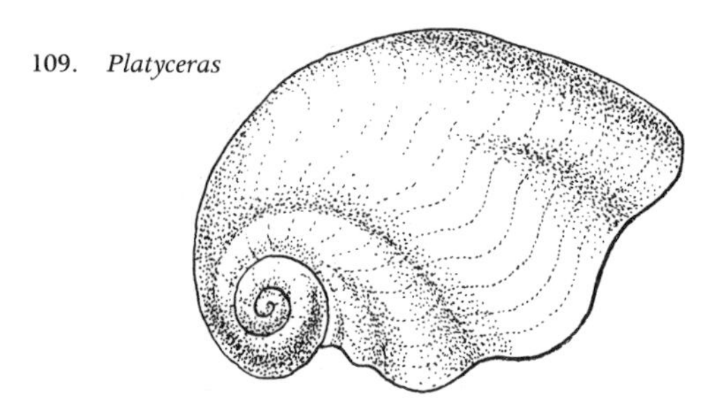

PLATYCERAS (Fig. 109)
Silurian to Permian. Shell loosely coiled; body whorl large, with weak transverse or spiral ridges and crenulated on aperture margin; aperture large, rounded. Up to $1\frac{1}{2}$ inches in diameter.

110. *Pleurotomaria*

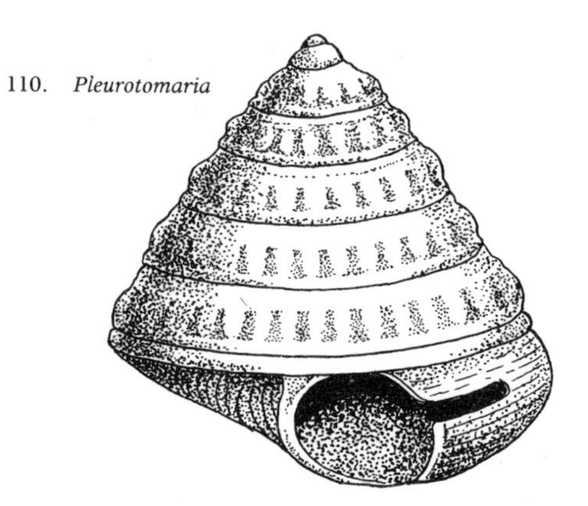

PLEUROTOMARIA (Fig. 110)
Jurassic-Lower Lias, to Cretaceous. Shell conical, like a Top Shell; whorls ornamented with numerous spiral ridges and ribs; aperture rounded; outer lip has a noticeable slit band. Up to $2\frac{1}{2}$ inches in height.

POLYGYRA
Palaeocene to Recent. Shell tightly coiled, flattened, but has large, rounded body whorl; smooth; aperture rounded, indented. Up to 1 inch in diameter.

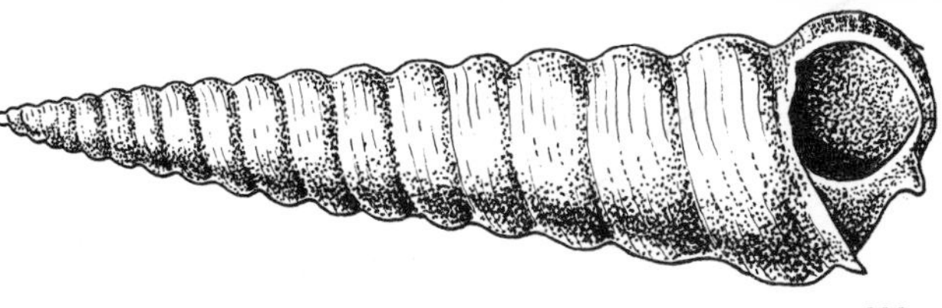

111. *Turritella*

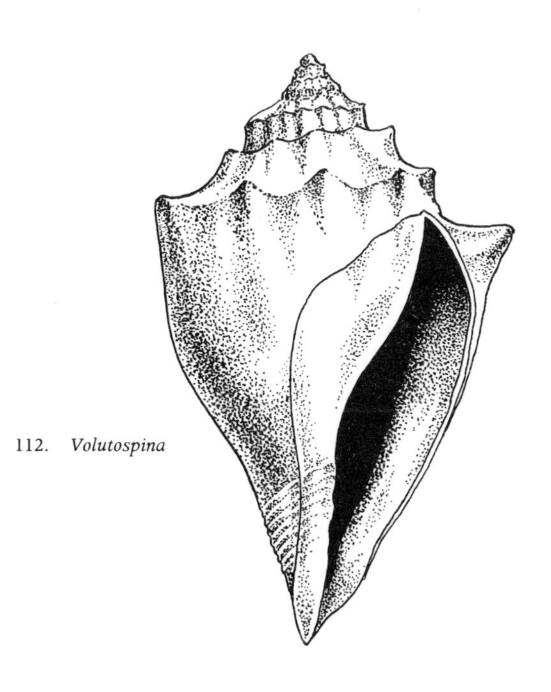

112. *Volutospina*

TURRITELLA (Fig. 111)
Cretaceous to Recent. Shell elongated, slender, with a tall spire; whorls rounded, with weak spiral ornamentation; aperture rounded. Up to 4 inches in height.

VIVIPARUS
Cretaceous-Weald Clay. Shell Snail-like; smooth, rounded whorls, the body whorl being the largest; spire conical; sutures deep; aperture large, rounded. Up to $1\frac{1}{2}$ inches in height.

VOLUTOSPINA [ATHLETA] (Fig. 112)
Eocene to Recent. Shell elongated, with a conical spire; whorls angulated, ribbed with blunt knobs or spines on the shoulders; body whorl very large; aperture large, elongated, narrow. Up to 5 inches in height.

Pelecypoda or Lamellibranchiata

The *Lamellibranchs*, 'plate-gilled' Molluscs, or *Pelecypods*, 'hatchet-foot' Molluscs, also exist today as a very large group of marine, limy-shelled, Bivalve Molluscs. Similar to the *Gastropoda* it is not possible in the confines of this book to describe all the various living forms and their anatomy, but this is entirely covered in Chapter 2 of my companion volume 'Collecting World Sea Shells' (Bartholomew). Fossil *Pelecypods* or *Lamellibranchs* occur from the Ordovician Period to Recent Period. Many of them, particularly those found in the Tertiary or Cainozoic Era rocks, such as *Astarte, Venus, Solen, Ensis, Ostrea, Pteria, Modiolus, Chama, Nucula,* etc., have present-day descendants. The valves, due to the decomposition of the ligament that held them together, may have become separated after death.

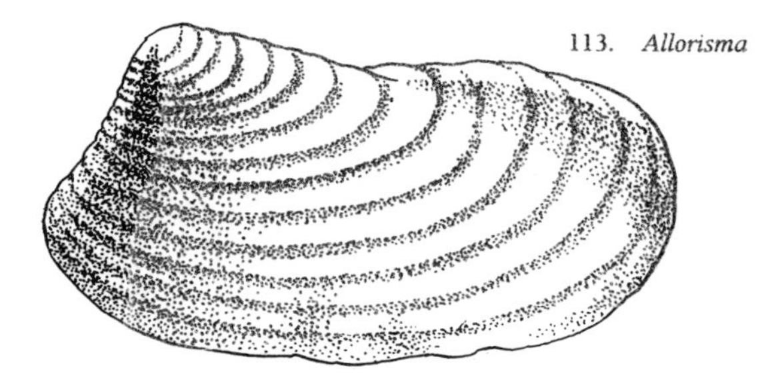
113. *Allorisma*

ALLORISMA [WILKINGIA] (Fig. 113)
Mississippian (North America), Carboniferous-Coal Measures (British Isles) and Permian. Shell elongated, oval; margin flat behind anterior beak; growth lines prominent. Up to $2\frac{1}{2}$ inches in length.

ASTARTE
Triassic to Recent. Shell triangular-rounded; beak prominent; smooth or has numerous concentric ribs; valve margins may be crenulated. Up to 1 inch in length.

AVICULOPECTEN
Silurian to Permian. Shell Scallop-shape, inequivalve; hinge straight; two 'ears' or 'wings'; numerous prominent ribs. Up to 1 inch in length.

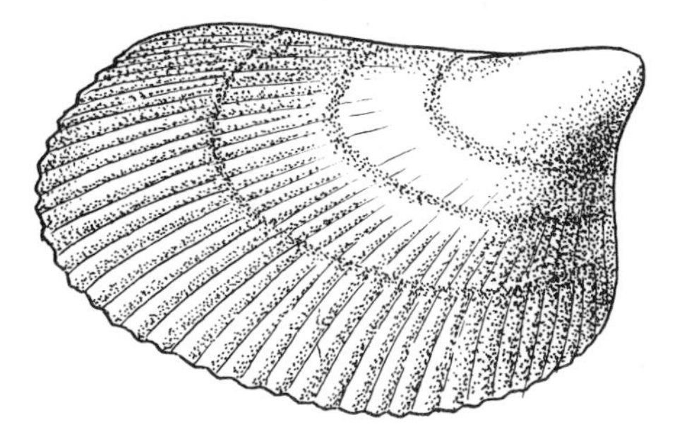

114. *Byssonychia*

BYSSONYCHIA (Fig. 114)
Ordovician. Shell triangular-rounded; equivalve; hinge straight; prominent inclining beak sited one end of hinge; prominent radial ribs. Up to 1 inch in length.

CARBONICOLA (Fig. 115)
Carboniferous-Coal Measures (British Isles), Pennsylvanian (North America). Shell elongated, oval; inequilateral; beak towards anterior margin; numerous concentric growth lines. Up to $1\frac{1}{2}$ inches in length.

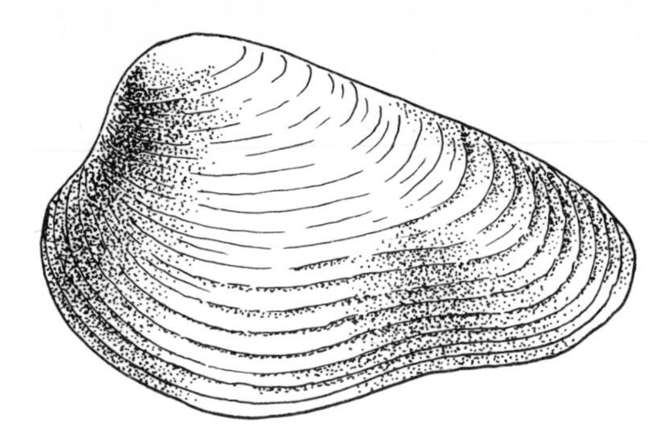

115. *Carbonicola*

CARDIUM
Triassic to Recent. Shell Cockle-shaped; equivalve; almost rounded; convex; beak curved; hinge curved; valve margins crenulated; prominent radial ribs. Up to 2 inches in length.

CHAMA
Eocene. Shell rounded; inequivalve, largest valve having a prominent curved umbo; prominent concentric growth lines. Up to 2 inches in diameter.

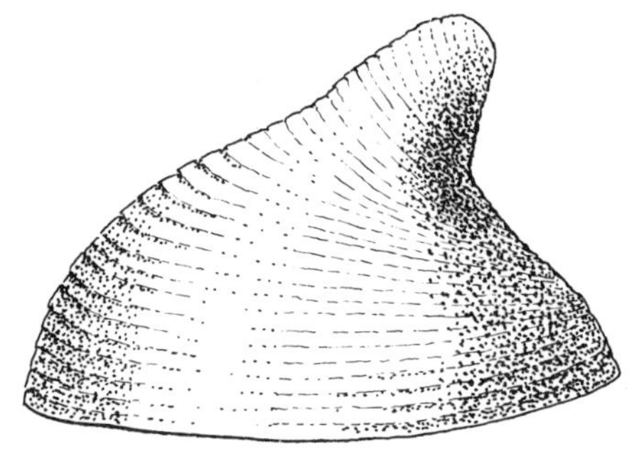

116. *Conocardium*

CONOCARDIUM (Fig. 116)
Ordovician to Permian. Shell triangular; inequivalve, posterior oblique, anterior short; appears twisted out of shape; hinge straight, long; beak prominent; numerous prominent radial ribs. Up to 2 inches in length.

CTENODONTA

Ordovician to Silurian. Shell triangular-rounded; equivalve; smooth, but may have weak concentric growth lines. Up to 1 inch in length.

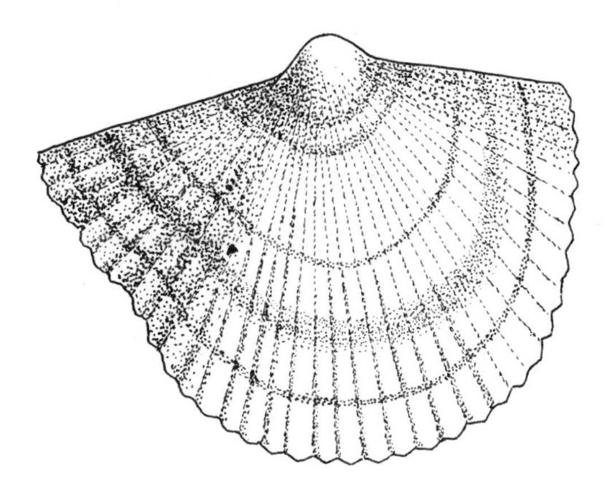

117. *Dunbarella*

DUNBARELLA (Fig. 117)

Carboniferous-Coal Measures, Millstone Grit (British Isles), Pennsylvanian (North America). Shell Scallop-shape; flat; hinge straight; the 'wings' or 'ears' being very wide and almost in alignment with rounded outer margin; numerous prominent radial ribs; weak concentric growth lines. Up to 2 inches in length.

ENSIS

Eocene to Recent. Shell Razor-shell shape (British Isles), [Jack-nife Clam (North America)]; flattened, margins straight or slightly curved, anterior and posterior ends truncated, tapered or rounded; weak concentric lines. Up to 9 inches in length.

EXOGYRA (Fig. 118)

Jurassic-Corallian Beds, to Cretaceous-Upper and Lower Green-sand. Shell approximately oval, but distorted by the umbones being twisted spirally; sometimes weak or prominent growth lines or ribs. Up to 5 inches in length.

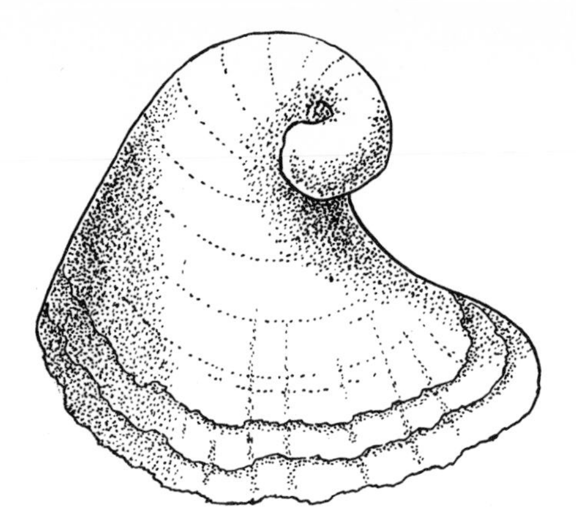

118. *Exogyra*

GLYCYMERIS

Cretaceous to Recent. Shell Dog-Cockle shape (British Isles), [Atlantic Bittersweet (North America)]; rounded; equivalve; beaks pointed, prominent, almost central on the broad, thick, toothed hinge; may have prominent or weak radial ribs and concentric growth lines. Up to $2\frac{1}{2}$ inches in diameter.

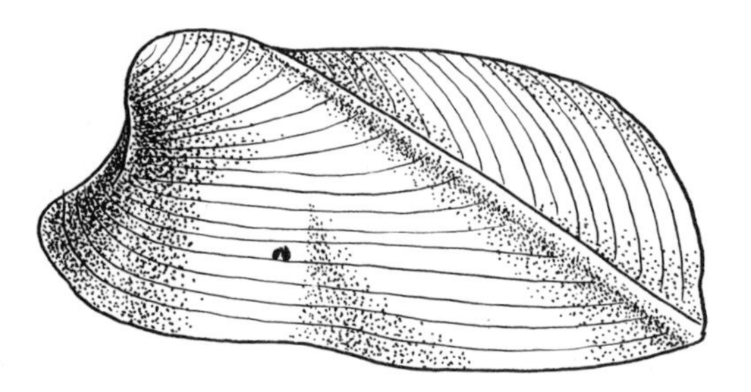

119. *Goniophora*

GONIOPHORA (Fig. 119)

Silurian and Devonian. Shell has a distorted appearance, with a prominent beak at the anterior margin and a prominent ridge extending from it to the posterior margin; weak concentric growth lines. Up to 2 inches in length.

120. *Grammysia*

GRAMMYSIA (Fig. 120)
Silurian to Mississippian (North America), Carboniferous (British Isles). Shell elongated-oval; beak incurving, prominent, blunt; has an oblique fold in the valves from the beaks to the valve margins; prominent concentric growth lines. Up to 2 inches in length.

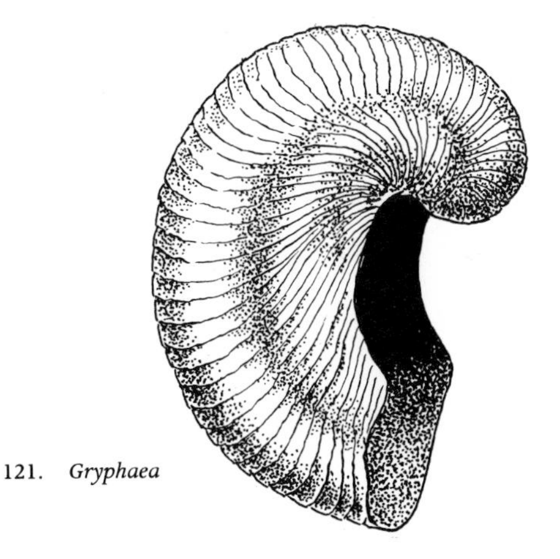

121. *Gryphaea*

GRYPHAEA (Fig. 121)
Jurassic-Lower Lias, to Eocene. Shell very inequivalve; a primitive Oyster, nicknamed 'the Devil's Toenails'; right valve flat like a lid against the much larger, loosely-coiled left valve; has numerous prominent growth lines. In some examples there may be almost no coiling. Not to be confused with *Exogyra*. Up to $3\frac{1}{2}$ inches in length.

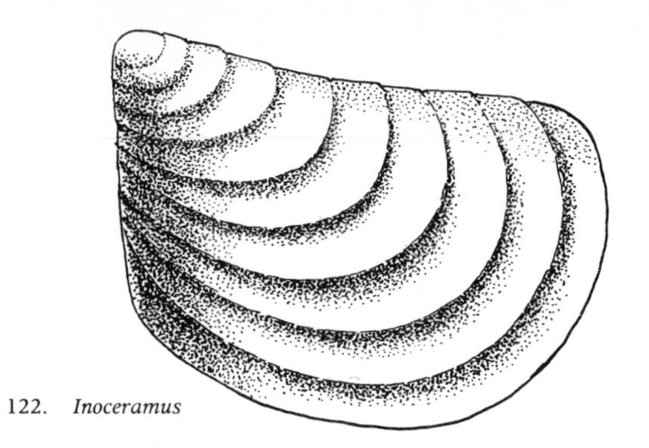

122. *Inoceramus*

INOCERAMUS (Fig. 122)
Jurassic to Cretaceous-Chalk, Gault. Shell oval; straight hinge; beak prominent; prominent concentric growth lines. Up to 4 feet in length.

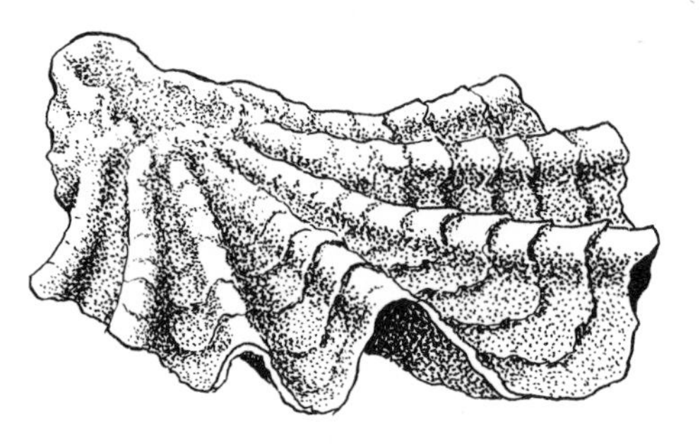

123. *Lopha*

LOPHA (Fig. 123)
Jurassic-Inferior Oolite, Corallian, to Cretaceous-Upper Chalk. Shell oval, irregular, with prominent plicated ridges or folds; folds of left and right valves intersect when closed. Up to $2\frac{1}{2}$ inches in length.

MERETRIX [PITAR] (Fig. 124)
Eocene-London Clay. Shell rounded; beak curved; fairly prominent, numerous growth lines. Up to 2 inches in diameter.

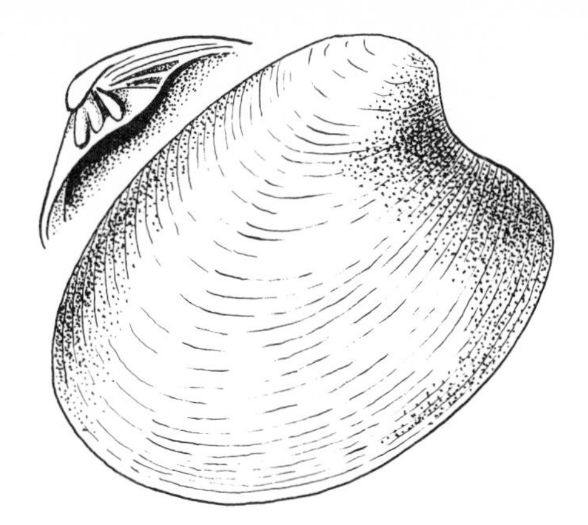

124. *Pitar*

MODIOLOPSIS
Ordovician, Silurian. Shell oval, thin; like a Mussel; valves asymmetrical with an oblique depression. Up to 1½ inches in length.

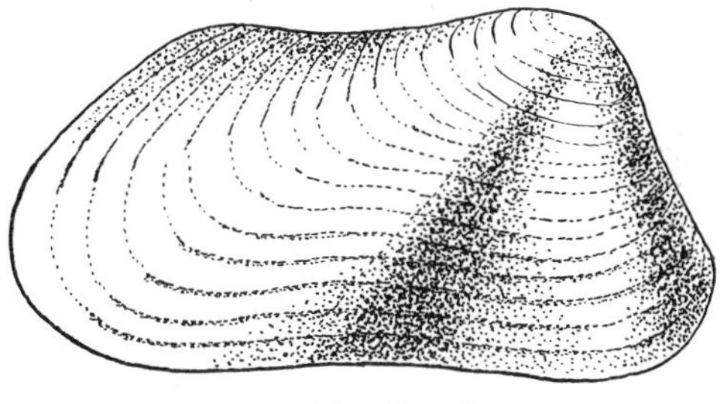

125. *Musculus*

MUSCULUS [MODIOLA] (Fig. 125)
Eocene-Barton Clay, London Clay. Shell oval, thin; shaped like a common Blue Mussel, but with numerous fine radiating ribs and concentric growth lines, although these may be weak in some examples. Up to 3 inches in length.

NUCULA
Silurian to Recent. Shell triangular-oval; prominent curved

beak; prominent or weak concentric growth lines. Up to 1½ inches in length.

OSTREA
Triassic to Recent. Shell elongated-oval or rounded; shaped like the Portuguese Oyster or Edible Oyster respectively; prominent irregular growth lines and deep folds; right valve flat and smaller than ribbed left valve. Up to 5 inches in length or diameter.

PANOPEA
Cretaceous-Lower Greensand, to Eocene-London Clay. Shell rounded-oval; prominent umbo; a burrowing Bivalve with posterior gape; numerous prominent or weak concentric growth lines. Up to 4 inches in length.

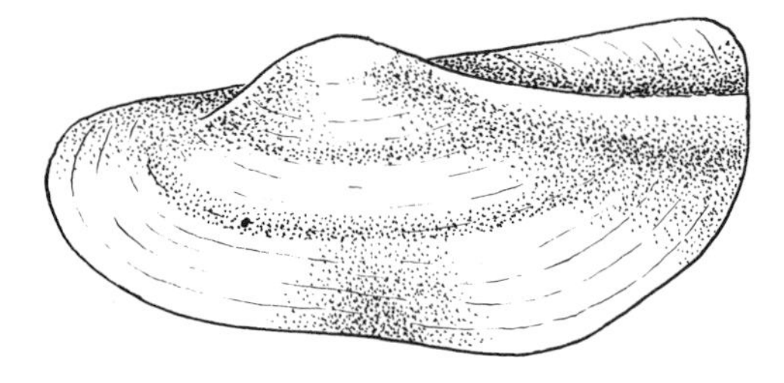

126. *Parallelodon*

PARALLELODON (Fig. 126)
Devonian to Jurassic-Great Oolite. Shell elongated; angular; hinge line long, straight; numerous concentric growth lines. Up to 1¼ inches in length.

PECTEN [ENTOLIUM]
Cretaceous-Upper Greensand. Shell circular; typical Scallop-shape; 'wings' or 'ears' equal size; numerous prominent, concentric growth lines. Up to 2½ inches in diameter.

PECTEN [NEITHEA]
Cretaceous-Upper Greensand. Shell circular; typical Scallop-shape; 'wings' or 'ears' inequal; hinge straight; numerous prominent radial ribs. Up to 4 inches in diameter.

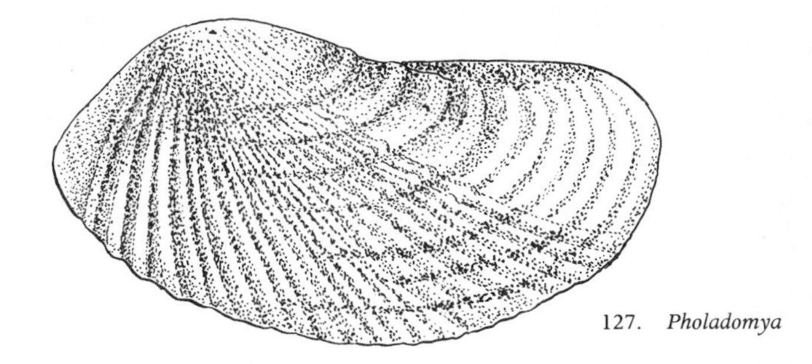

127. *Pholadomya*

PHOLADOMYA (Fig. 127)
Jurassic-Inferior Oolite. Shell rounded-oval; a burrowing Bivalve similar to the American Piddock or Fallen Angel Wing; numerous prominent radiating ribs and concentric ridges or growth lines; curving beak at anterior end. Up to 3 inches in length.

PINNA
Jurassic to Recent. Shell triangular, wedge-shaped, like a half-closed fan; similar to the Fan Mussels (British Isles) and Pen Shells or Sea Pens (North America); due to original thin valves may be fossilized as broken fragments; posterior end may be curved; beak forward at narrow, anterior end; hinge long; numerous, sometimes weak, radial ridges and concentric growth lines. Up to 10 inches in length, less broad.

PTERIA
Jurassic to Recent. Shell valves inequilateral; shaped like the Wing Oysters; posterior 'wing', rear of the beak which is curved forward towards anterior end; in front of the beak there is a smaller, triangular anterior 'wing'; hinge long, straight; may have radial ribs and weaker concentric growth lines. Up to 3 inches in length.

PTERINEA [PALAEOPECTEN] (Fig. 128)
Silurian-Wenlock, Ludlow, to Pennsylvanian (North America), Carboniferous (British Isles). Shell similar to a Scallop-shape except is less rounded and has a pronounced curve; opposite valves dissimilar; 'wings' or 'ears' inequal; hinge wide, straight;

numerous prominent radial ribs and weaker concentric growth lines. Up to 1½ inches in length.

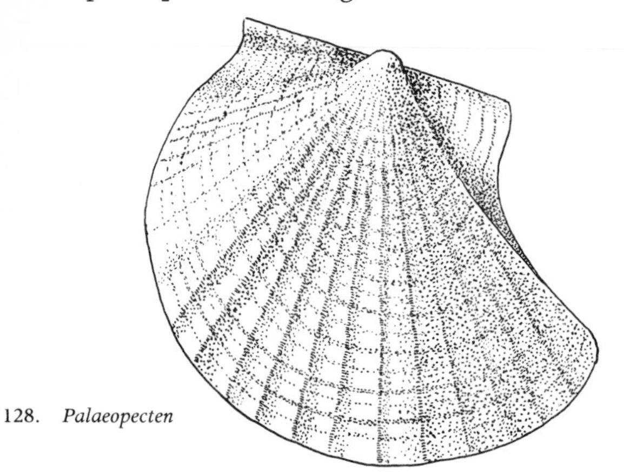

128. *Palaeopecten*

TRIGONIA

Jurassic to Recent. Shell triangular or crescent-like shape, with a prominent ridge from beak to margin; beak points backwards; noticeable different ornamentation on anterior and posterior sides of valves; prominent curving concentric growth lines sometimes broken into nodules and irregular, noduled and curved or straight growth lines. Up to 3½ inches in length.

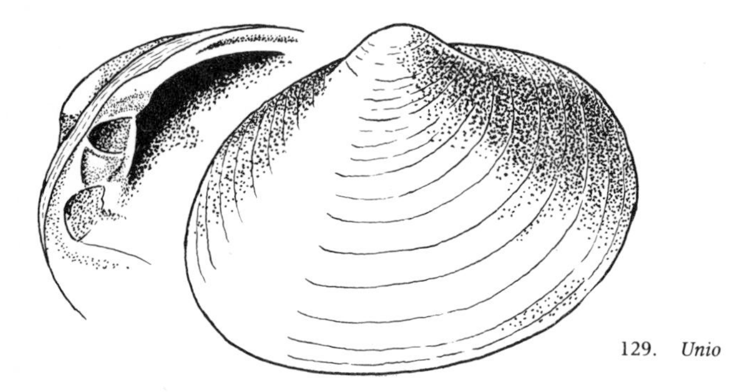

129. *Unio*

UNIO (Fig. 129)

Triassic to Recent. Shell oval; equivalve; similar to Freshwater Mussel or Clam; beak prominent, curved, blunt; concentric growth lines weak; due to original frailty of thin shell may be crushed or fragmented as a fossil. Up to 3 inches in length.

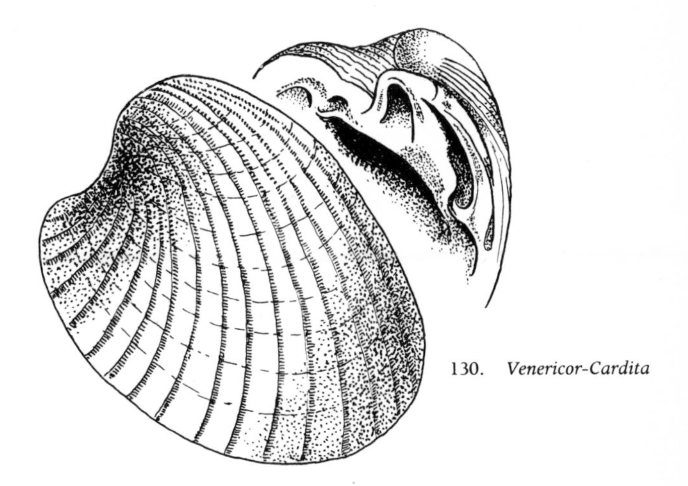

130. *Venericor-Cardita*

VENERICOR [CARDITA] (Fig. 130)
Eocene. Shell rounded; similar to Cockle-shape; umbo or beak prominent, curved forward; numerous ribs radiating from beak to margin; numerous concentric lines. Up to 3 inches in length.

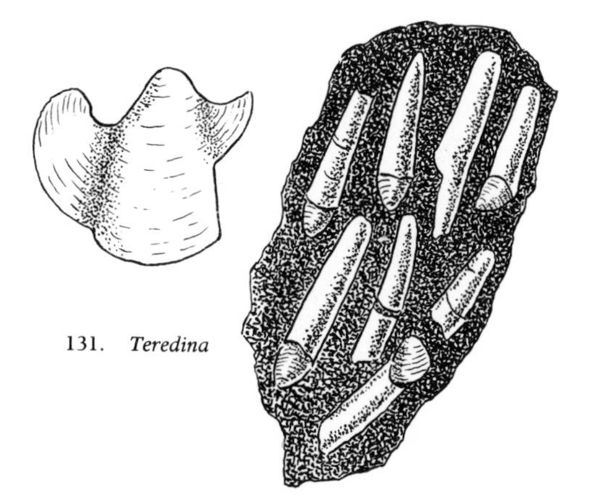

131. *Teredina*

(Fig. 131)

MOLLUSCS AND BORINGS (See WORMS AND BORINGS)

Some of the *Lamellibranchs,* such as *Pholadomya, Panopea, Lithophaga, Solen, Ensis,* etc., burrowed into sand, silt, soft mud or soft rock, as part of their existence and their fossilized burrows, with, in some cases, the fossil *Lamellibranch* therein or close by, also occur. Fossil wood, which was for a time drifting in the sea, occurs frequently in marine clays of the Mesozoic and Tertiary Eras with profuse borings of the *Teredo* or Shipworm, one example being *Teredina,* found in wood in the Eocene-London Clay. Within the burrows there may also be the small shell valves used as the boring instrument.

BRACHIOPODA

The *Brachiopods,* 'arm-foot', or Lamp Shells, live today as small marine *Invertebrates,* one example being *Lingula,* but the number of present species now in hundreds is drastically lower than the thousands abundantly discovered as fossils. Although they have a similarity to *Pelecypod Bivalves,* with a bilateral symmetry, the *Brachiopods* have two inequal sized valves. They are divided into two groups, the *Inarticulata* and the *Articulata.* The valves of the *Inarticulata* group, comprising shells of chitin and calcium phosphate, are kept together by muscle contraction only. The larger *Articulata* group, with limy shells, also have a hinge apparatus, with teeth on the ventral or pedicle, larger, valve, which fit into sockets on the dorsal or brachial, smaller, valve. Fossil *Brachiopods* occur from the Cambrian Period to the Recent. The fossil Cambrian *Brachiopods* are mostly *Inarticulate* species and less frequently discovered than the later abundantly common *Articulate* species, the latter being particularly those of the Ordovician and Silurian Periods.

Inarticulate

LINGULA (Fig. 132)

Ordovician to Recent. Shell thin, oval or tear shaped; valves' anterior margin rounded or straight; posterior margin pointed;

shape of both valves being nearly identical; weak concentric growth lines; may be found fossilized in its original burrow. Up to 1½ inches in length.

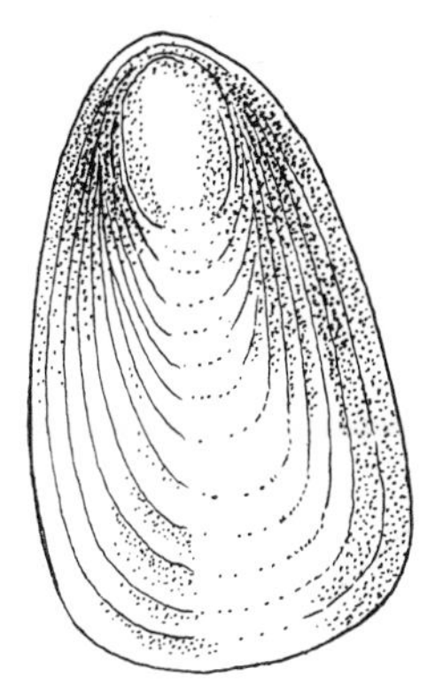

132. *Lingula*

LINGUELLA (Fig. 133)

Cambrian to Ordovician. Shell pointed-oval or tear shaped; valves' anterior margin rounded; posterior margin pointed; fairly prominent concentric growth lines. Up to 1 inch in length.

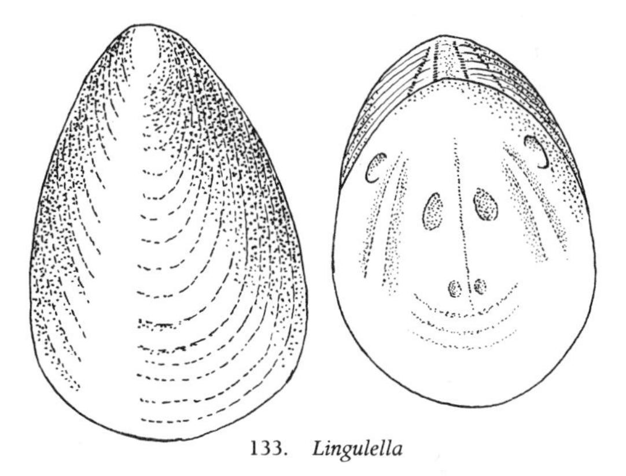

133. *Lingulella*

OBOLELLA (Fig. 134)

Lower Cambrian, Ordovician. Shell triangular-rounded or almost circular, except for pointed posterior margin; concentric growth lines fairly prominent or weak. Up to ½ inch in length.

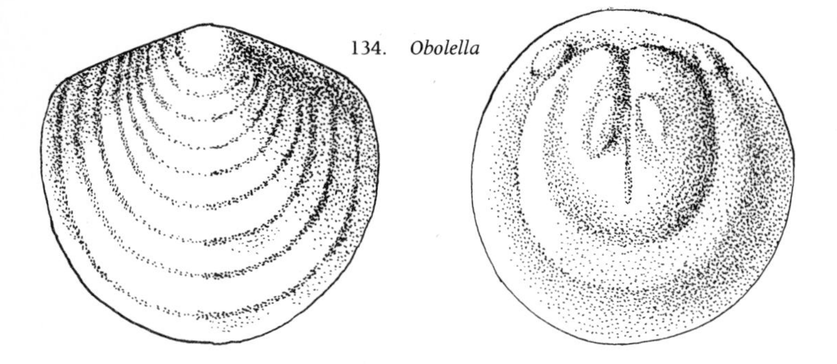

134. *Obolella*

ORBICULOIDEA
Ordovician to Permian. Shell rounded; brachial valve conical; the umbo being near to posterior margin; weak concentric growth lines. Up to ¾ inch in diameter.

Articulate

ATRYPA (Fig. 135)
Silurian-Wenlock, to Lower Mississippian (North America), Devonian (British Isles). Shell rounded; brachial valve convex; ventral valve slightly convex; numerous radiating ribs and concentric lines. Up to 1¼ inches in length.

135. *Atrypa*

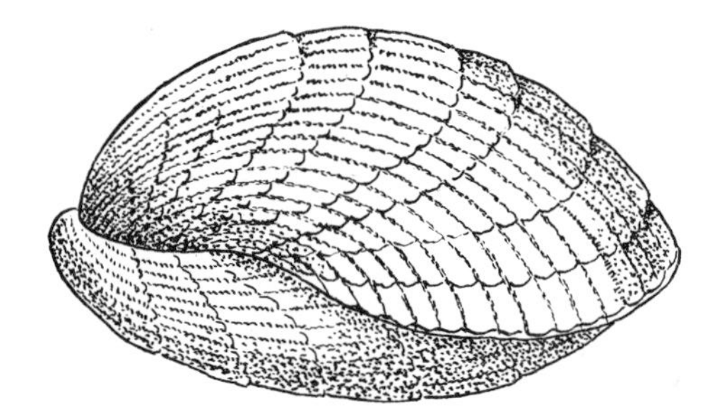

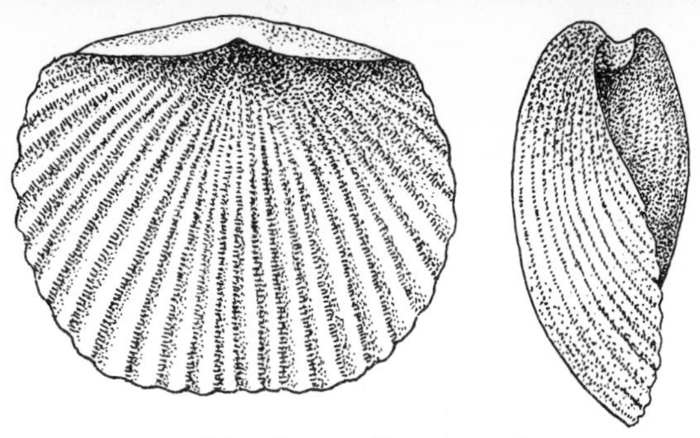

136. *Chonetes (Protochonetes)*

CHONETES [PROTOCHONETES] (Fig. 136)
Silurian-Ludlow. Shell semi-circular; brachial valve concave; hinge wide, with a varying number of small spines on margin; numerous weak radiating ribs and several concentric growth lines. Up to 1 inch in length.

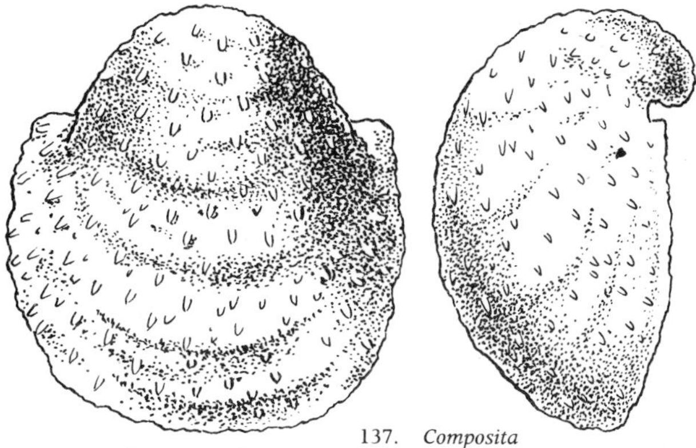

137. *Composita*

COMPOSITA (Fig. 137)
Mississippian (North America), Carboniferous (British Isles) to Permian. Shell sub-circular, convex; pedicle valve has a sulcus; weak concentric growth lines. Up to 1 inch in length.

CONCHIDIUM
Silurian-Ludlow. Shell oval, convex; prominent radial ribs; weak growth lines. Up to 1 inch in length.

DINORTHIS [ORTHIS] (Fig. 138)
Ordovician-Caradoc. Shell rounded-oval; bi-convex; hinge straight; prominent radial ribs. The *Orthids* are a large group of common fossils. Up to 1 inch in length.

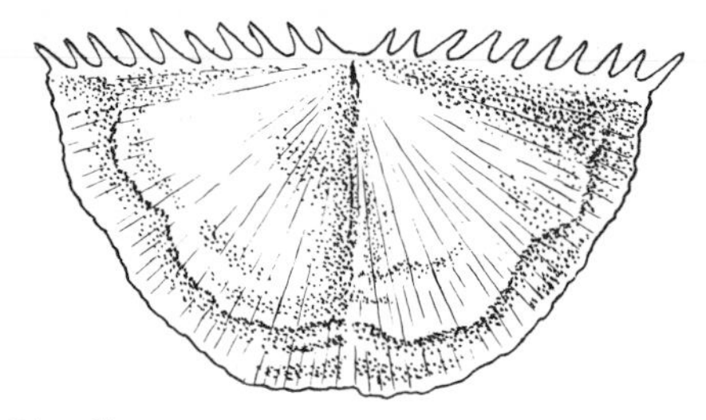

138. *Dinorthis*

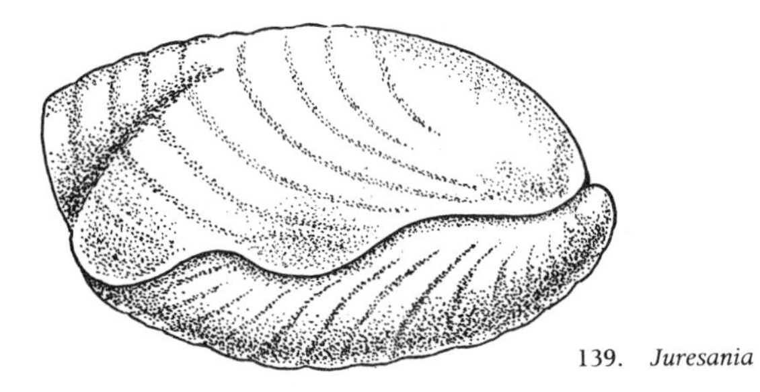

139. *Juresania*

JURESANIA (Fig. 139)
Pennsylvanian (North America) to Permian. Shell semi-circular; brachial valve concave, pedicle valve very convex; hinge straight; brachial and pedicle valves have short spines or remains of spines. Up to $1\frac{1}{2}$ inches in length.

LEPIDOCYCLUS [RHYNCOTREMA]
Ordovician. Shell sub-circular, convex; fold in brachial valve; hinge short; prominent radial ribs; sometimes has 'herring-bone' pattern on valves. Up to $1\frac{1}{4}$ inches in length.

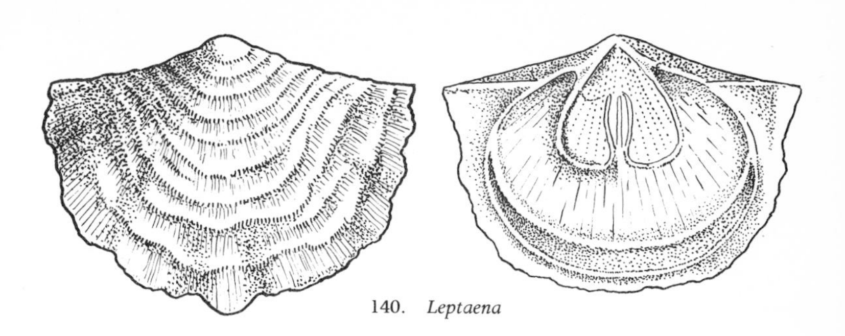
140. *Leptaena*

LEPTAENA (Fig. 140)
Ordovician to Carboniferous (British Isles), Mississippian (North America). Shell semi-circular; brachial valve concave, pedicle valve convex; valves irregular along margins; hinge near-straight; weak radiating ribs, more prominent concentric growth lines. Up to $1\frac{1}{2}$ inches in length.

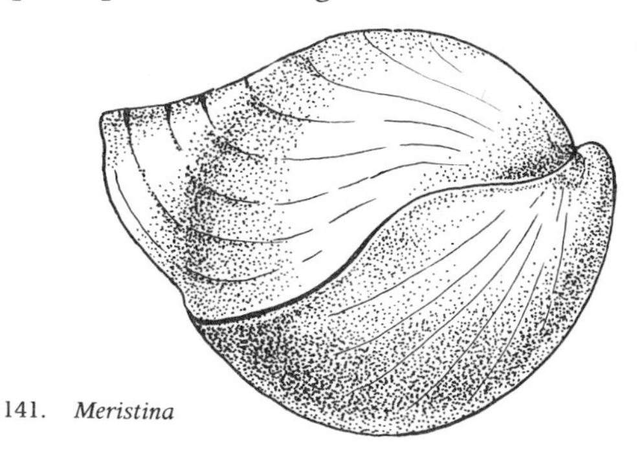
141. *Meristina*

MERISTINA (Fig. 141)
Silurian-Wenlock, Ludlow. Shell rounded, convex; fold in brachial valve; valves almost smooth with weak radial ribs and concentric growth lines. Up to 1 inch in length.

PENTAMERUS (Fig. 142)
Silurian-Llandovery. Shell rounded-oval, convex; pedicle valve has vertical plate; valves smooth, weak growth lines. Up to 3 inches in length.

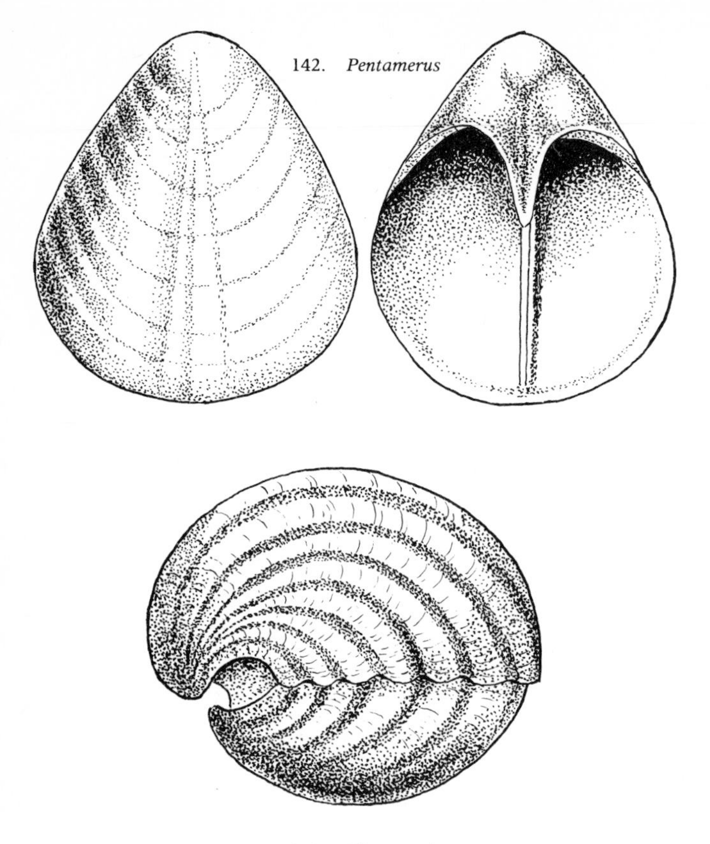

142. *Pentamerus*

143. *Platystrophia*

PLATYSTROPHIA (Fig. 143)
Ordovician to Silurian. Shell sub-circular, convex; hinge straight; prominent radial ribs. Up to $1\frac{1}{2}$ inches in length.

PRODUCTUS (Fig. 144) (Fig. 145)
Carboniferous (British Isles). Shell semi-circular; convex ventral or pedicle valve, concave dorsal or brachial valve; hinge straight; umbones incurving; numerous radial ribs and concentric growth lines. Up to 1 inch in length. Two related species are: GIGANTO-PRODUCTUS, Carboniferous (British Isles). Shell semi-circular; wide; with 'wing'- or 'ear'-like extensions; hinge straight; prominent ribs. Up to 2 feet in width. LINOPRODUCTUS,

Carboniferous (British Isles), Mississippian (North America) to Permian. Shell triangular; long, convex pedicle or ventral valve, may be wrinkled in hinge area; umbo incurving; prominent ribs; may have spines or spine base remains on ventral valve. Up to 1 inch in length.

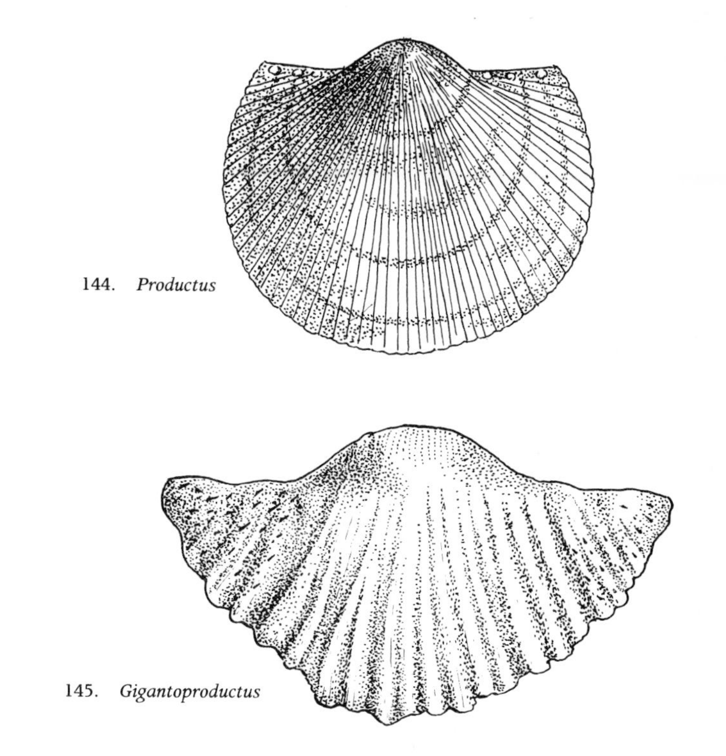

144. *Productus*

145. *Gigantoproductus*

RHYNCHONELLA (Fig. 146)
Ordovician to Recent. Shell triangular, convex; hinge short; may have a sulcus; prominent ribs, concentric growth lines. The *Rhynchonellids* are a large group of common fossils. Up to $1\frac{1}{2}$ inches in length.

SPIRIFER (Fig. 147) (Fig. 148) (Fig. 149)
Devonian (British Isles), Mississippian (North America) to Pennsylvanian (North America), Carboniferous (British Isles). Shell rounded-triangular, convex; a prominent sulcus or groove;

hinge wide; prominent radial ribs. Up to 1 inch in length. The *Spiriferid Brachiopods* are a large group of common, variable fossils. Other examples are: NEOSPIRIFER, Pennsylvanian (North America), Carboniferous (British Isles) to Permian. Shell triangular, wide, convex; hinge straight, prominent; variable, prominent fold; prominent radial ribs. Up to $1\frac{1}{4}$ inches in length. MUCROSPIRIFER, Devonian. Shell triangular, but has a very extended hinge; prominent radial ribs; prominent fold. Up to 1 inch in length. EOSPIRIFER, Silurian to Devonian. Shell sub-oval or semi-circular, convex; hinge straight; wide fold or sulcus; weak, fine radial ribs. Up to $1\frac{1}{4}$ inches in length.

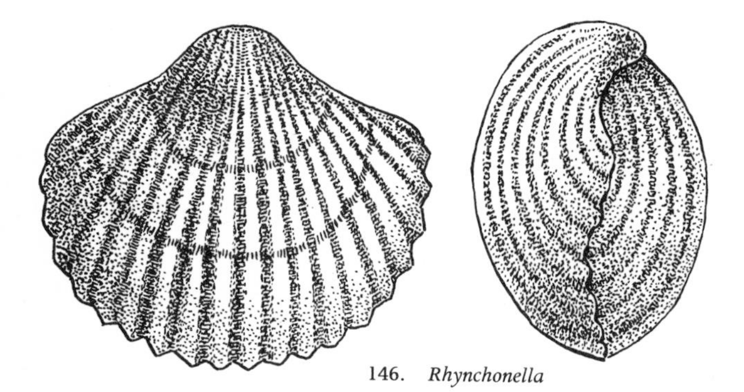

146. *Rhynchonella*

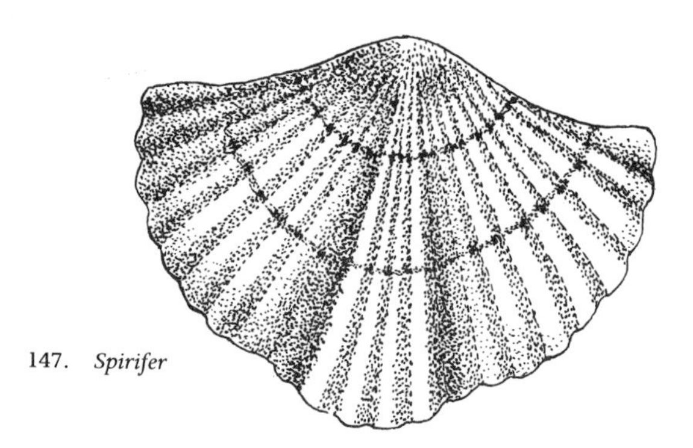

147. *Spirifer*

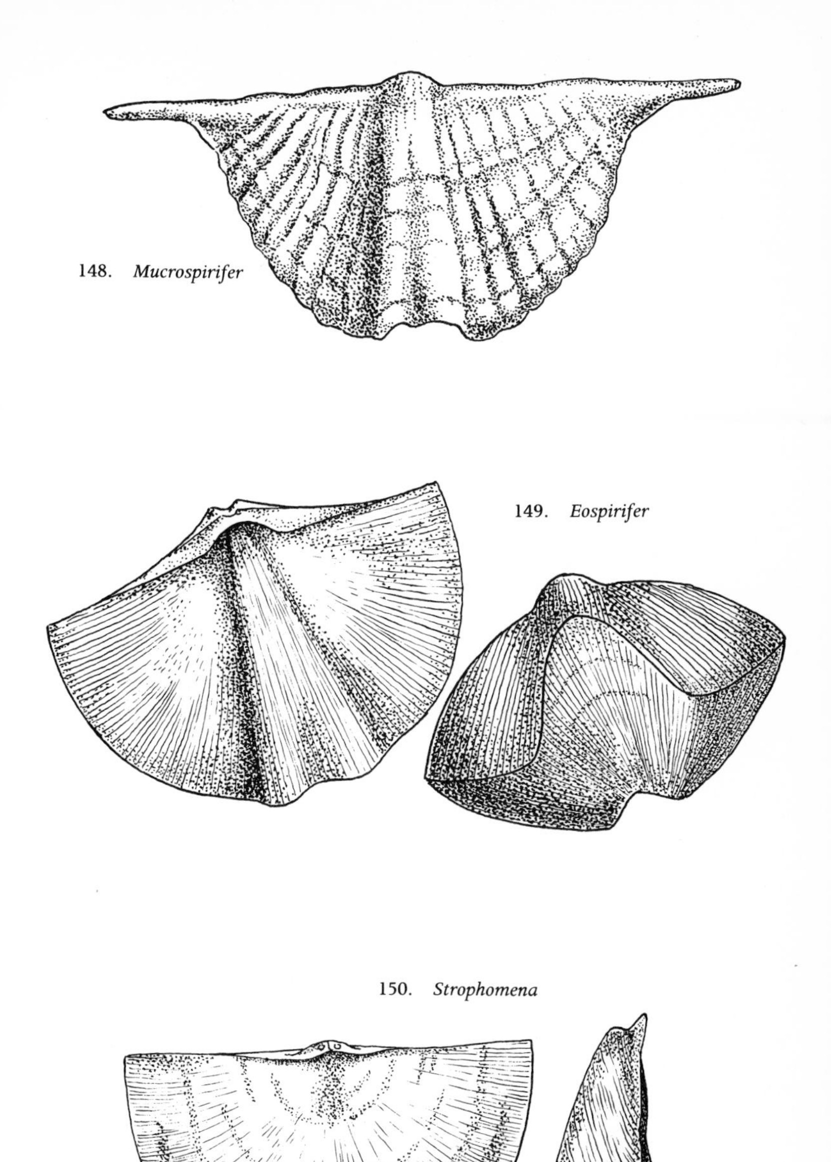

148. *Mucrospirifer*

149. *Eospirifer*

150. *Strophomena*

STROPHOMENA (Fig. 150)
Ordovician. Shell semi-circular; brachial valve convex, pedicle valve concave; hinge long, straight; fine radial ribs. Up to $1\frac{1}{2}$ inches in length.

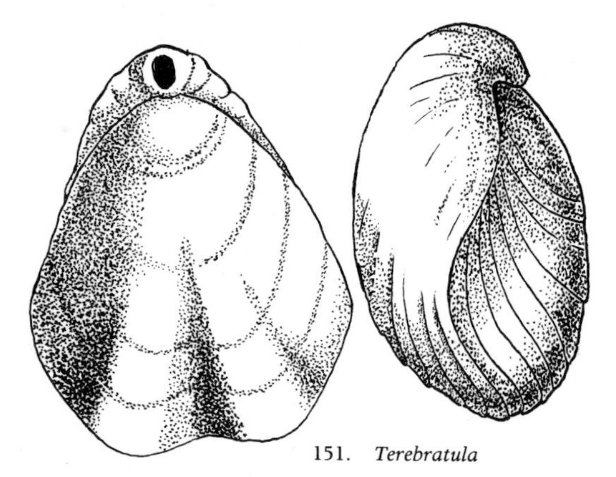
151. *Terebratula*

TEREBRATULA (Fig. 151)
Silurian to Pleistocene. Shell oval, convex; may have variable waved folding; numerous weak concentric growth lines. Up to 4 inches in length, but more usually less.

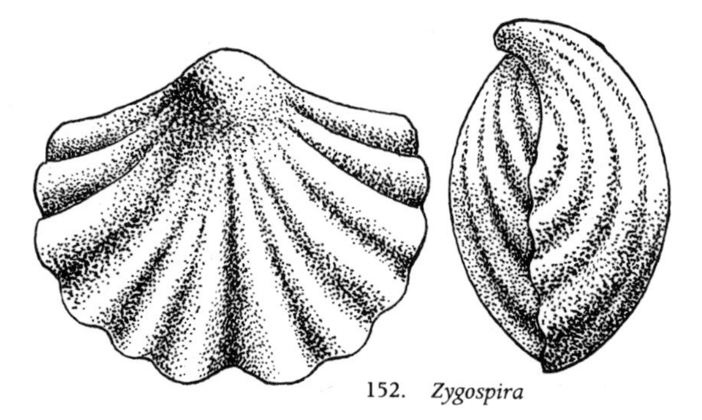
152. *Zygospira*

ZYGOSPIRA (Fig. 152)
Ordovician, Silurian. Shell semi-circular, bi-convex; pedicle valve has a sulcus; prominent radial ribs. Up to $\frac{1}{2}$ inch in length.

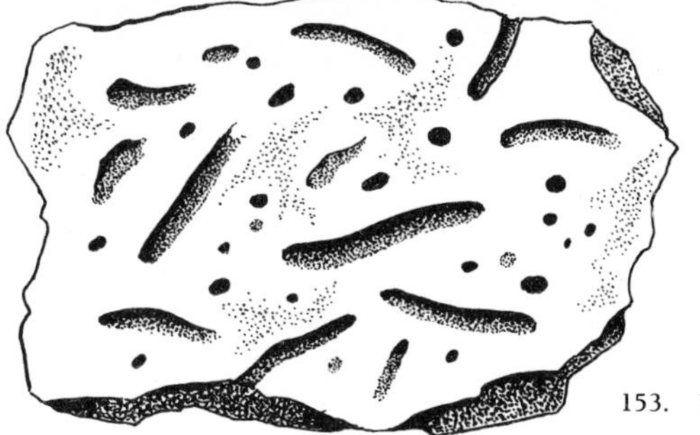

153. Worm Borings

WORMS AND BORINGS (See Molluscs and Borings) (Fig. 153)
As is well known, various groups of worms exist today, but the only fossil Worms are the *Annelida* or Segmented Worms. They are scarce, dating back to the late pre-Cambrian Era, but impressions of Worm bodies have been discovered, principally in North America and Australia.

The toothed-jaw parts of marine Annelid Worms, very tiny, chitinous, or siliceous, structures, or *Scolecodonts*, occur as micro-fossils from the Ordovician to Recent Period. Much more common and easier to see and obtain are fossil Worm borings, burrows, tubes, casts or trails (See Trace Fossils) in sedimentary

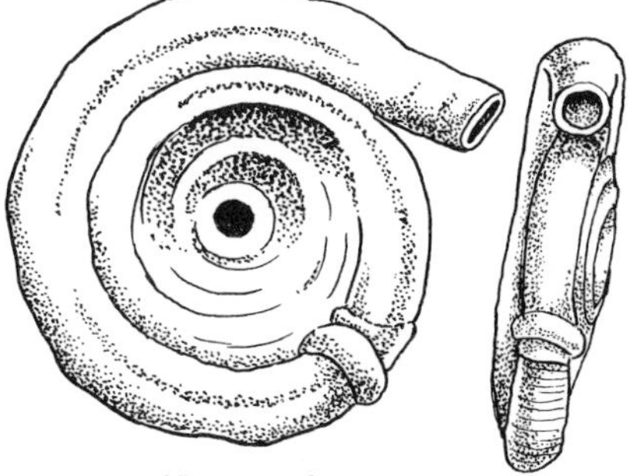

154. *Rotularia*

(Fig. 154) (Fig. 155) (Fig. 156) (Fig. 157) (Fig. 158) (Fig. 159) rocks. Some of the Worms, like many present-day species, created with body secretions, a limy tube or supporting lining,

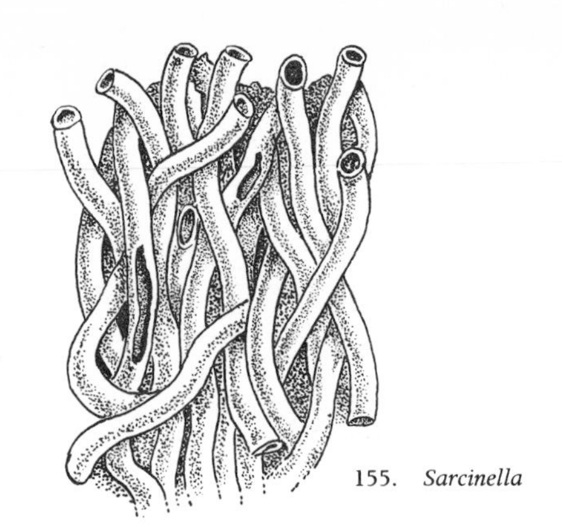

155. *Sarcinella*

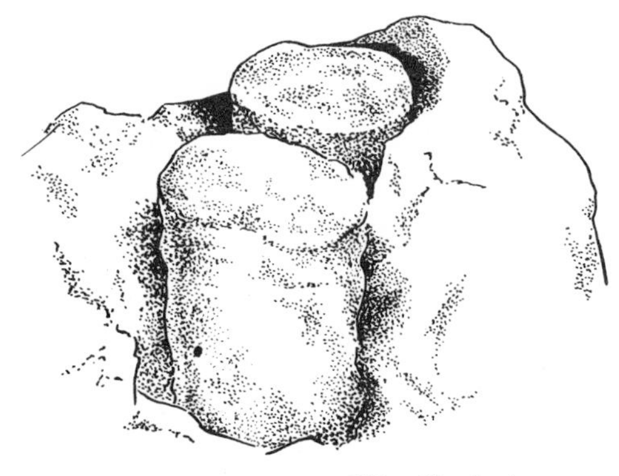

156. Pipe Rock

157 *Tentaculites*

in which to exist, the tube becoming fossilized. One example is the irregular, loosely- or tightly-coiled, limy tubes of *Serpula*, Silurian to Recent, up to 1 inch in length. Another is *Spirorbis*, Silurian to Recent, which created very small, coiled tubes, round in cross-section. *Rotularia*, Cretaceous, Eocene, had small, tightly-coiled tubes, round in cross-section, which sometimes occur in colonies. The straighter tubes of *Sarcinella*, Cretaceous, several inches in length, may be found in intertwining groups.

Further evidence that Worms existed in the Cambrian Period is the 'pipe-rock' which occurs in north-west Scotland. Originally this was a sand of the Lower Cambrian Series, but now is a quartzite, and contains a profusion of the original Worm burrows which give it its name. These vertical burrows are up to 2 inches in diameter, straight or curving and the remains found are several inches in length, so the Worms that lived therein must have been considerably long and broad.

In addition to the true Worm fossil remains and evidence there are extinct forms discovered which may have been primitive Worms. One of these is *Tentaculites*, Ordovician to Devonian, which is several inches in length, tapering, and loosely spiralled, resembling a long-threaded screw. A similar form is *Cornulites*, Ordovician to Silurian, which is the same length, but thicker with the spirals closer together. A third uncertain form is *Conularia*, Ordovician to Carboniferous (British Isles), which is up to 6 inches in length, sharply pyramidal in shape, quadrilateral in cross-section, with a pattern of close lines.

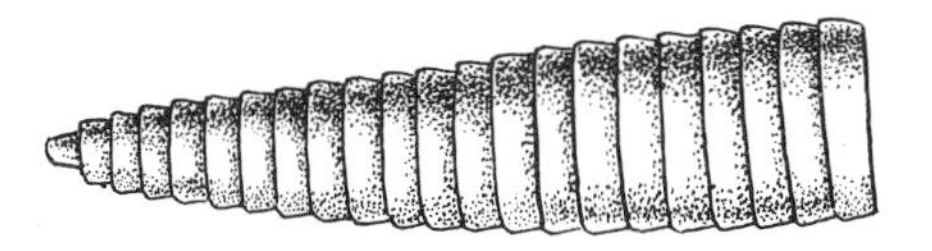

158. *Cornulites*

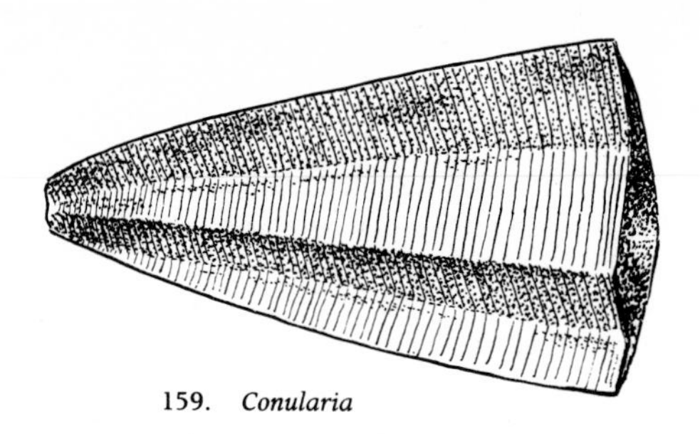
159. *Conularia*

TRACE FOSSILS

(Fig. 160)

Trace Fossils are the evidence that an organism walked or crawled on, or burrowed into, the sediment or rock while this was still soft. In doing so it left behind its tracks which hardened and may have later been filled with another sediment, silt or mud, which preserved them. Footprints and trails of the Reptiles and Amphibians are one example. Another is the impression marks made by the five 'arms' of Starfish. Straight, intertwining or U-shaped tubes may have been created by Worms, similar to those of the Lugworm today. Care should be taken not to confuse these genuine examples of organism movements with PSEUDO-FOSSILS.

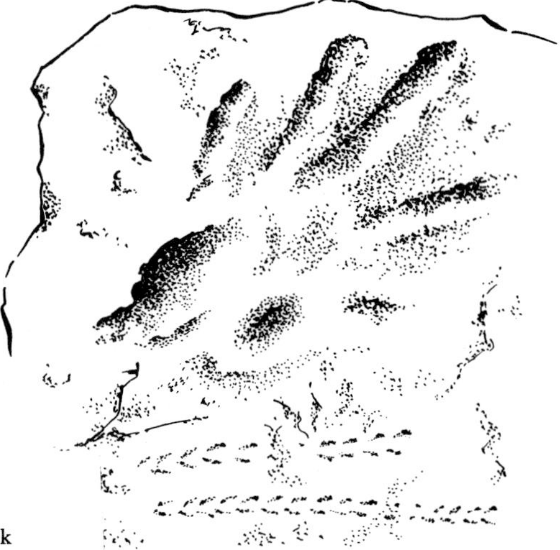
160. Trace fossils—
Dinosaur footprint above
below, unknown double track

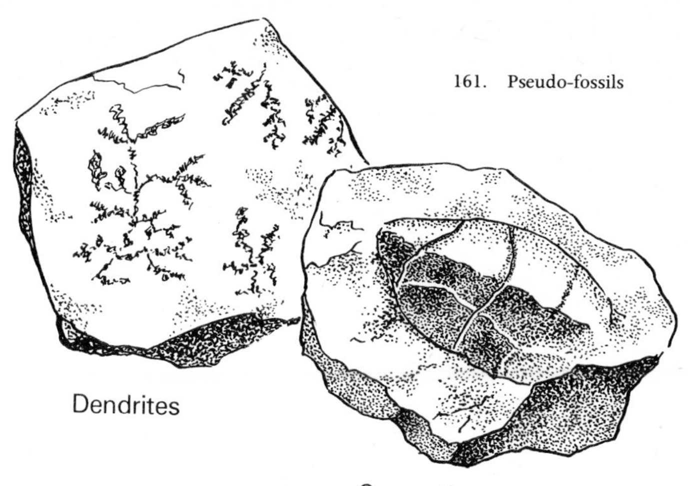

161. Pseudo-fossils

PSEUDO-FOSSILS

(Fig. 161)

Pseudo-fossils are, as their name indicates, rock shapes which at first glance externally resemble fossils. They may be formed by water dissolving mineral in one position to be deposited elsewhere as concretions. Sometimes they are created around another organism, in sedimentary rock, and may contain a fossil, but only infrequently. Flints may be in a shape very similar to *Belemnites*. Concretions of limestone can look externally like fossil Coral. Rain and wind may also have eroded a depression so that it appears to be a fossil footprint in the rock. Another misleading instance is that of *Dendrites* or fern-like deposits created on or within certain rocks. Although they appear to be fossil plants, like moss, they are in reality mineral in origin. The experienced fossil-collector eventually learns to recognise *Pseudo-fossils* and to identify them with the help of data as to the localities where they are found. For example, 'footprints' may be in rock which is known to have formed millions of years before Reptiles, Mammals or Amphibians could have walked on it and left their tracks. Microscopic examination of *Pseudo-fossils* also reveals they do not have the internal detailed cellular structure of the true fossils. But if you are in real doubt do not throw the specimen away before letting a geologist, an experienced fossil-collector, or your local geology museum have a look at the object to confirm its identity.

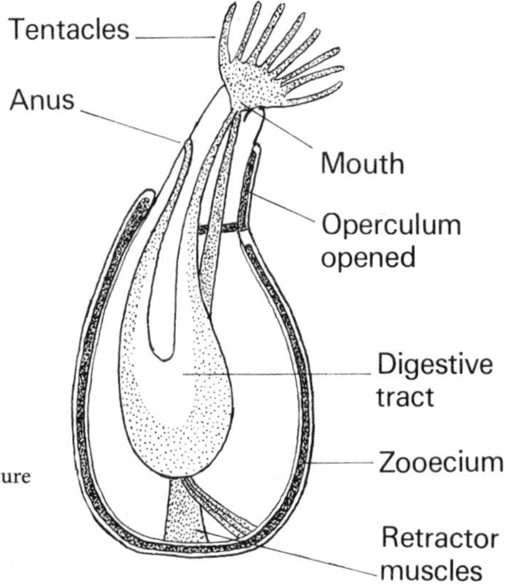

162. *Bryozoa* structure

POLYZOA OR BRYOZOA

(Fig. 162)
The *Polyzoa*, 'many-animals', or *Bryozoa*, 'moss-animals', can be found on the lower seashore, living as marine colonial organisms, forming horny or limy encrustations, lace-like areas, or short 'sticks', on rocks and Seaweeds, or creating a branched, Seaweed-like colony. One example of the latter is the familiar Sea-Mat or Horn Wrack. Several also occur in freshwater. Under a lens it will be seen that the *Polyzoans* or *Bryozoans* consist of numerous, small, separate compartments, each containing a tiny individual animal. Fossil *Polyzoans* or *Bryozoans* occur from the Ordovician to Recent Periods, particularly in limestones and crags. As they are sometimes found encrusted on shells of *Brachiopods* and *Pelecypods,* also tests of *Echinoids,* close exam-

ination should be given to these as a possibility of discovering *Polyzoans*.

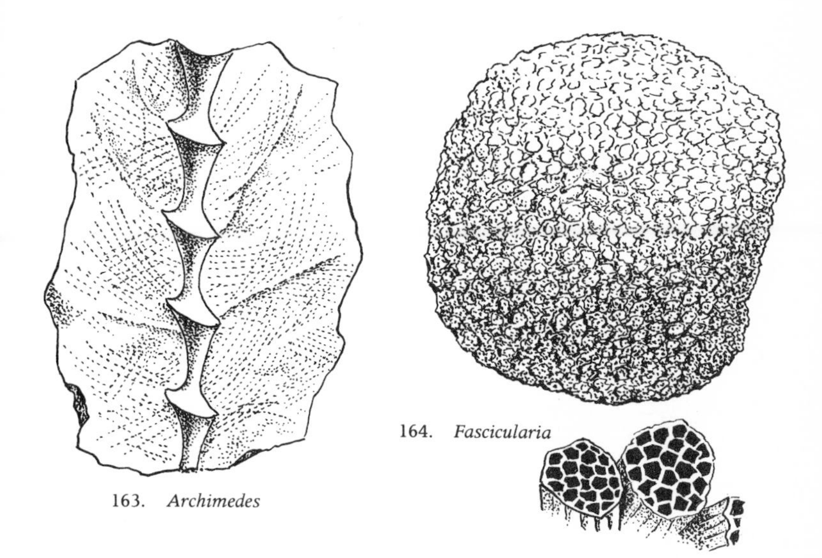

163. *Archimedes*

164. *Fascicularia*

ARCHIMEDES (Fig. 163)
Mississippian (North America), Carboniferous (British Isles) to Permian. Colony has a corkscrew-like axis with lace-like branches. Up to 2 inches in length.

FASCICULARIA [MEANDROPORA] (Fig. 164)
Pliocene-Coralline Crag. Colony circular encrustation, with varying numbers of individuals forming separate, irregular, short branches of compartments.

FAVOSITELLA (Fig. 165)
Silurian-Wenlock. Colony honeycomb-like encrustation; area being variable in size.

FENESTELLA (Fig. 166)
Carboniferous-Limestone, to Permian-Magnesian Limestone. Colony net- or lace-like encrustation; area being variable in size.

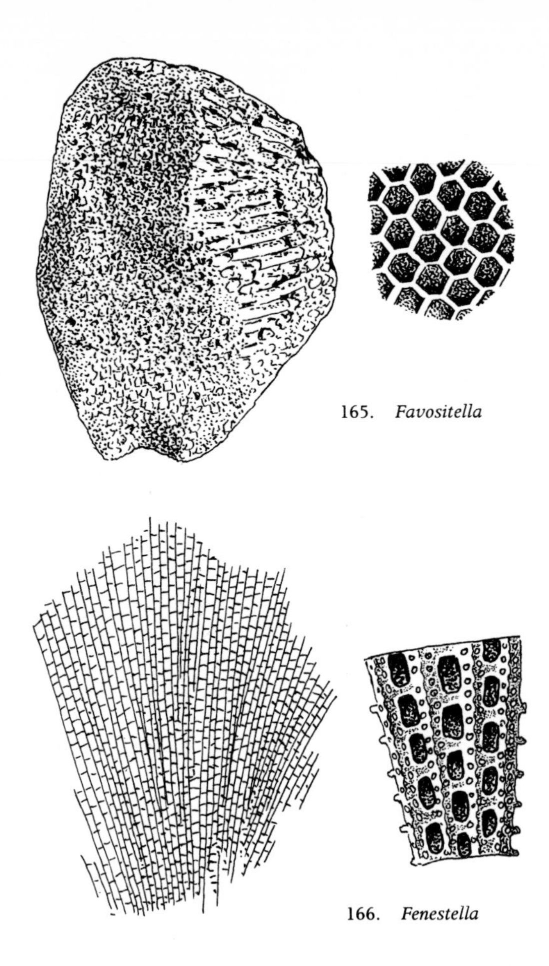

165. *Favositella*

166. *Fenestella*

RHOMBOPORA

Ordovician to Permian. Colony narrow branches; may be spined; variable in length.

COELENTERATA

The *Coelenterata* contains the familiar present-day marine Jellyfish, Sea-Anemones, Corals, Hydroids or Sea Firs, etc. They are simple in construction, typically have a radial symmetry, but the exterior appearance is dissimilar. Some resemble flowers or when in an elaborate colony are like a stalked plant. They have a sac-like body, with a mouth-anus and gastric cavity, from which was derived their name 'coel', Greek for hollow, and 'enteron', Greek for gut. The entire anatomy and life story of each class is related in my book 'Coast, Estuary & Seashore Life' (Gifford). They are divided into three classes–the *Hydrozoa, Scyphozoa* and *Anthozoa*. Fossils of the *Coelenterata* occur from the Pre-Cambrian to Recent Periods.

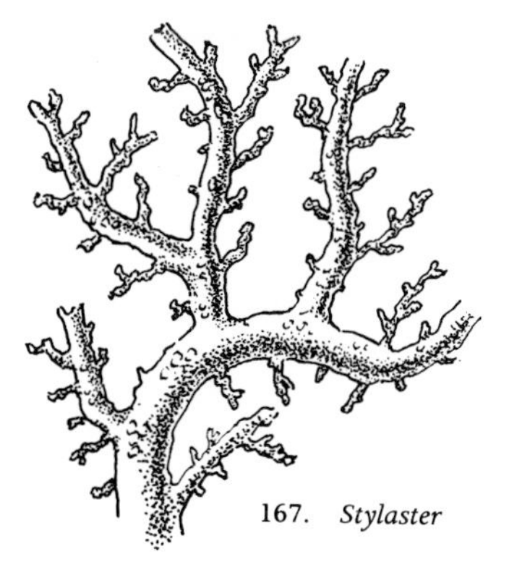

167. *Stylaster*

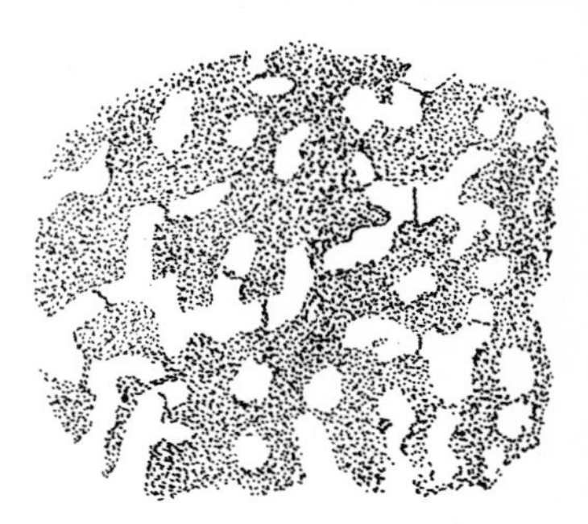

(Fig. 167)

Hydrozoa

The *Hydrozoa* or Sea Firs, which live in the sea, attached to rocks or Seaweeds, are composed of minute horny cups on stalks. Fossil *Hydrozoans* are rare, due to their soft structure. One example, however, *Stylaster,* in the sub-class *Stylasterida,* occurring as a fossil in the Eocene to Recent, is a colonial *Hydrozoan,* with numerous, narrow, limy branches, about 3 inches in length, and is part of the construction of present-day Coral reefs.

(Fig. 168)

The *Stromatoporoids* are a now-extinct group of marine colonial *Coelenterates,* with a limy skeleton, which occur fossilized as rounded, spherical or branching masses or thin encrusting layers. Externally they may have numerous pore-like holes and internally have numerous thin layers. Each of these 'pores' originally contained a living *Hydroid. Stromatopora*, Ordovician to Permian, has numerous, small, pore-like holes on its exterior, these being the apertures of limy tubes. In cross-section a colony comprises a mass of these tubes laminated in thin layers. The *Stromatoporoids* occur from the Cambrian to the Cretaceous, particularly in the Silurian and Devonian Periods, as important limestone rock builders and occur with Rugose and Tabulate Corals, also Algae. Some authorities class the *Stromatoporoids* with the Corals, others with the *Hydrozoans* and I have followed the latter course because the *Stromatoporoids* are more primitive than the higher Corals.

Scyphozoa

The *Scyphozoans* or Jellyfish, being soft-bodied and having no hard parts are very rarely preserved as fossils. Where Jellyfish rested or were stranded, an impression was fossilized and some of these impressions, up to 16 inches in diameter, are approximately the same size as present-day species. Fossil Jellyfish occur from the Cambrian Period.

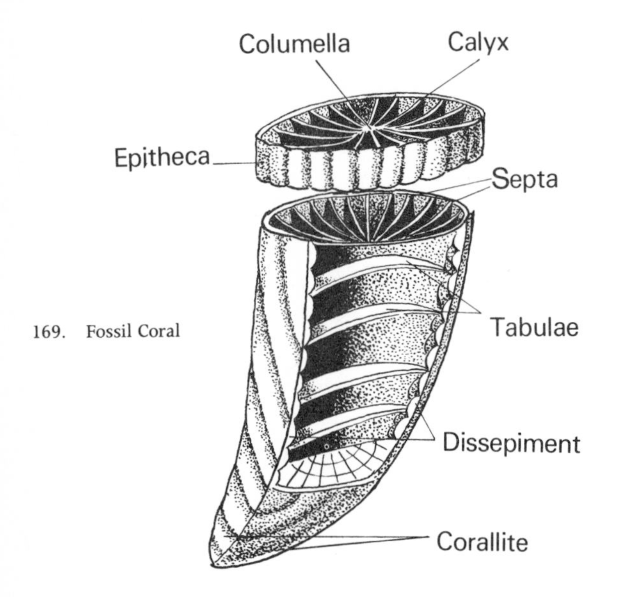

169. Fossil Coral

(Fig. 169) **Anthozoa**

The *Anthozoans* or Corals are more likely to be familiar to and recognised by the fossil hunter than the other two previous classes of the *Coelenterata.* There are three main classes of Corals–the *Octocorals* or *Alcyonaria,* the *Tabulata* and the *Zoantharia.* It should not be thought that fossil Corals occur only in tropical, warm, shallow water areas of the world as living reef Corals do today. Fossil Coral reefs have been discovered in Siberia, Canada, the U.S.A., as well as in the British Isles, proving that at the time these reefs were created the climate must have been very different from today. In addition to reef Corals there were also solitary Corals. Fossil Corals occurred from the Ordovician to Recent.

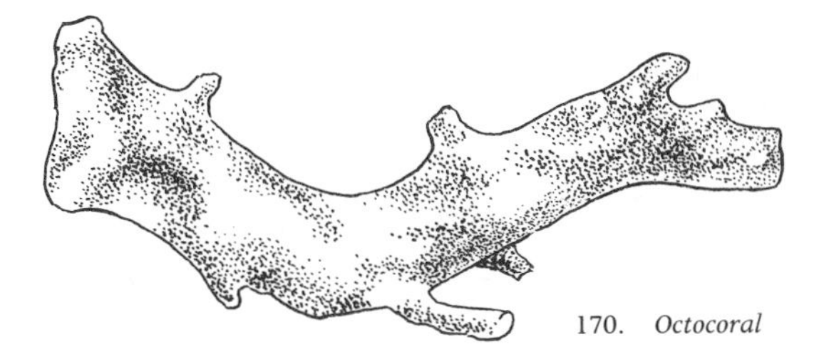

170. *Octocoral*

(Fig. 170) **Octocorals or Alcyonarians**

These Corals are one of the two present-day sub-classes of living limy or horny Corals. The colonial *Octocorals* also occur as fossils from the Triassic to Recent. Two examples are *Tubipora,* the Organ-Pipe Coral, and *Precious Coral.*

Tabulata

The *Tabulate Corals,* now extinct, were colonial, compound Corals, comprising a limy skeleton of numbers of individual cell-tubes or 'corallites' and strongly-developed, close-spaced 'tabulae' or horizontal partitions. Some of them had weak, or no, 'septa'-thin walls, radiating from the centre of a saucer-like hollow. There was no axial rod or columella. The earliest examples of *Tabulate Corals* occur in Lower Ordovician rocks in the U.S.A. In the British Isles and Europe they occur from the

Ordovician to the Jurassic Periods. The colony size may vary from a few inches in length, at its base, increasing upwards to a diameter of several inches or over a yard at its upper surface.

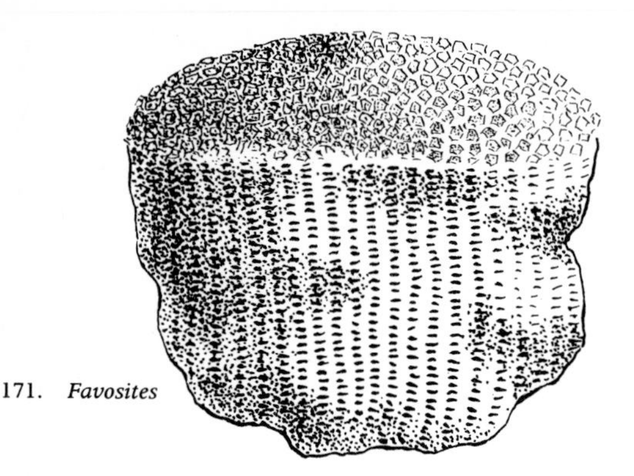

171. *Favosites*

FAVOSITES (Fig. 171)
Upper Ordovician to Upper Devonian. A 'honeycomb' Coral; the numerous small cells are approximately the size of a Bee's honeycomb cells, hence its name; the corallites are parallel and close together; in cross-section polygonal; no septa.

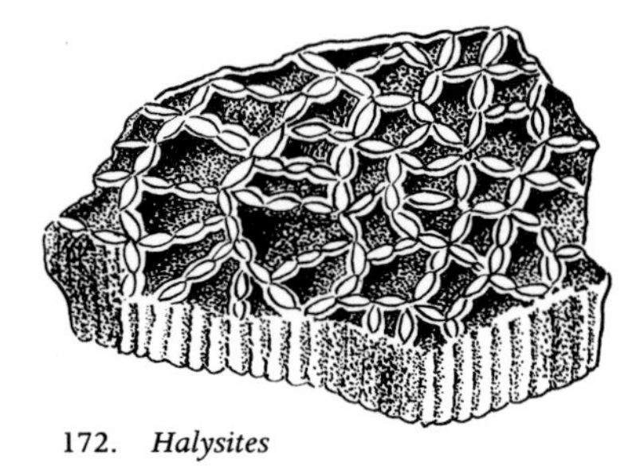

172. *Halysites*

HALYSITES (Fig. 172)
Upper Ordovician to Lower Devonian. A 'chain' Coral; the slender corallites are sited in branching rows so their thecae, seen upon the surface of the Coral cushion, resemble links in a chain, hence its name; in cross-section elliptical or circular; septa weak or absent; strong tabulae.

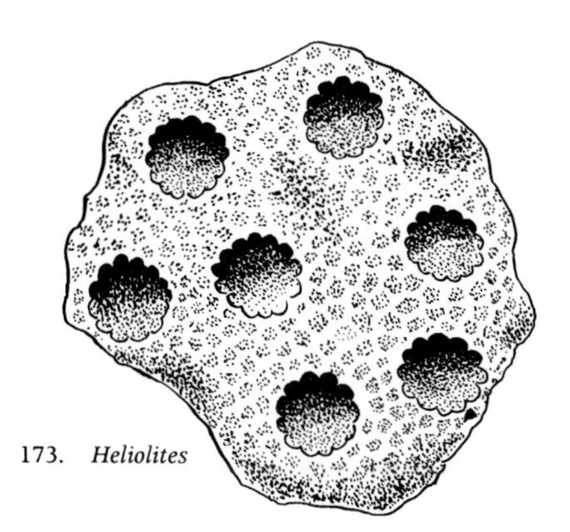
173. *Heliolites*

HELIOLITES (Fig. 173)
Silurian to Middle Devonian. Similar in shape to *Favosites*; has numerous cells of two sizes; the larger circular cells have short septa around their edges; the round corallites are separated from each other by the fine-grained, honeycomb-like network of smaller cells.

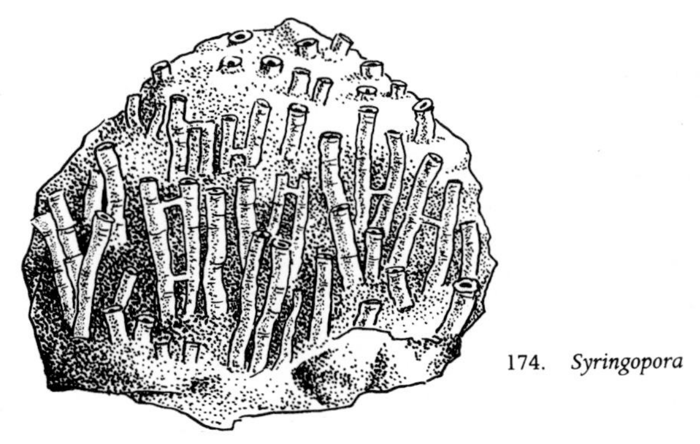
174. *Syringopora*

SYRINGOPORA (Fig. 174)
Silurian to Pennsylvanian (North America), Carboniferous (British Isles). A compound Coral; the corallites are clearly prominent with transverse or intersecting connections; septa spine-like; funnel-shaped tabulae.

Zoantharia

The *Zoantharian Corals* have a limy skeleton comprising a basal disc with an 'epitheca' or outer wall, and vertical 'septa' between the 'mesentaries,' the 'tabulae' extending in arches across the cup or 'theca'. The outer walls are sometimes supported by curved dissepiments and the hollow cup or 'theca' may have a central axial column.

There are two groups of *Zoantharian Corals*–the *Rugose* (Wrinkled) or *Tetracorals,* and the *Hexacorals* or *Scleractinia.* The solitary Corals are also known as 'Horn' Corals.

RUGOSE OR TETRACORALS

Some of the *Rugose Corals* have the main septa developed in four quadrants and so are called *Tetracorals.* The first *Rugose Corals* occur in Ordovician rocks in the U.S.A. and are similar to the first *Tabulate Corals.* They are particularly common from the Silurian to the Permian Periods, but became extinct at the beginning of the Mesozoic Era. The *Rugose Corals* are very variable and some of them, although they occurred in different Periods, closely resemble each other, so that study of the shape or arrangement of their structure, plus recognition of the Period rock is necessary for accurate identification.

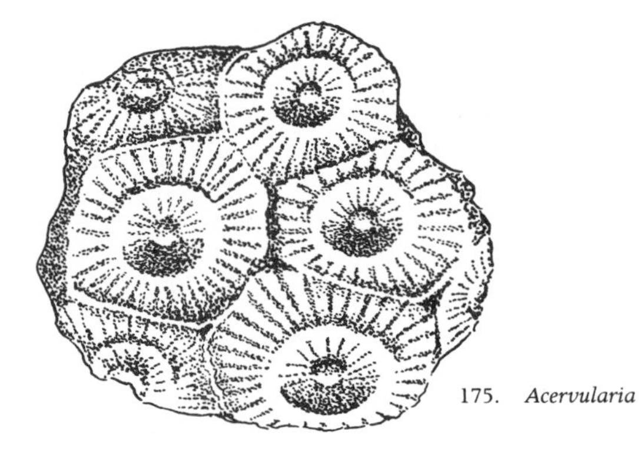

175. *Acervularia*

ACERVULARIA (Fig. 175)

Silurian, Devonian. A compound Coral; corallites polygonal; septa thicker from midway along their length creating an inner wall inside the thecal wall; has a pattern daisy-like in appearance over its surface.

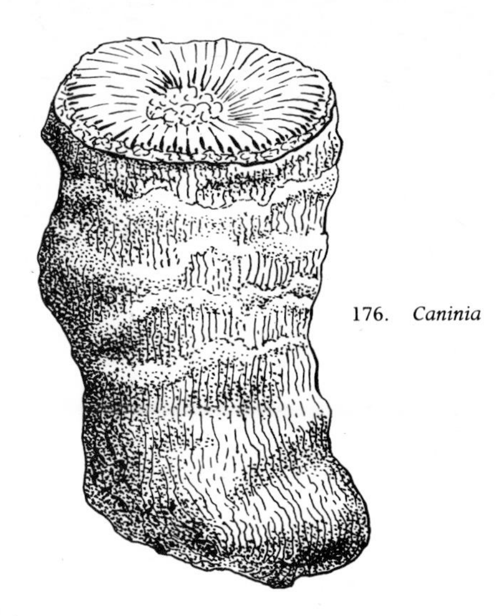

176. *Caninia*

CANINIA (Fig. 176)
Carboniferous. A solitary Tetracoral; cylindrical, sometimes curving and shaped like a cornucopia; septa radial; large specimens up to 12 inches in length may be found.

CYATHOPHYLLUM
Silurian, Devonian. A solitary Coral; corallite cylindrical; septa numerous, but only extend partly towards corallite's centre.

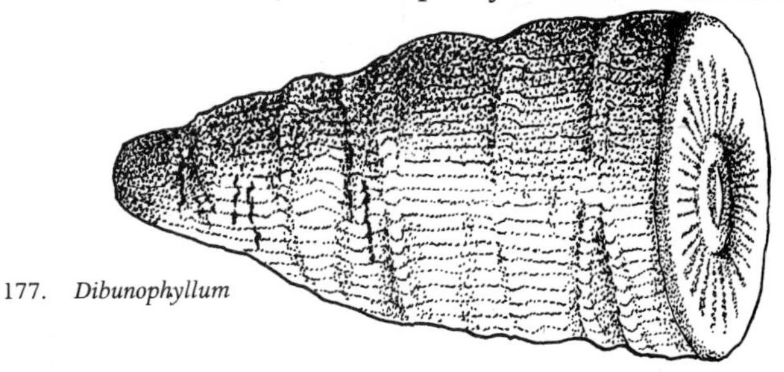

177. *Dibunophyllum*

DIBUNOPHYLLUM (Fig. 177)
Carboniferous (British Isles). A solitary Coral; corallite cylindrical; septa long and radially form a spider-web pattern, because their inner and outer ends are enclosed in a zone of vesicles; central axial column clearly defined.

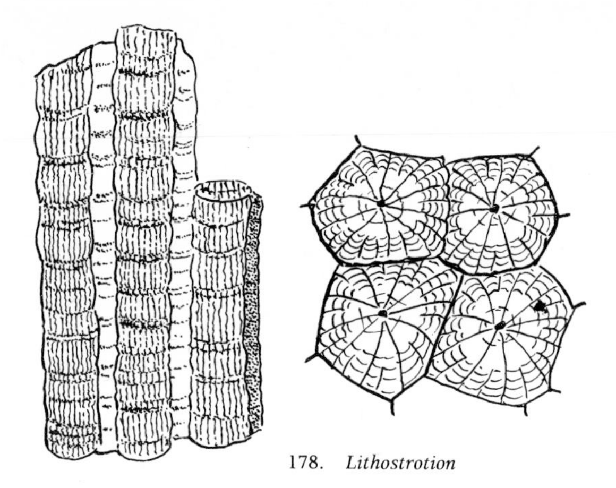

178. *Lithostrotion*

LITHOSTROTION (Fig. 178)
Carboniferous (British Isles), Mississippian to Pennsylvanian (North America). A compound Coral; corallites cylindrical or prismatic; few septa prominent extending radially towards but not as far as the central flattened axial rod or columella; tabulae central; has peripheral and columella dissepiments.

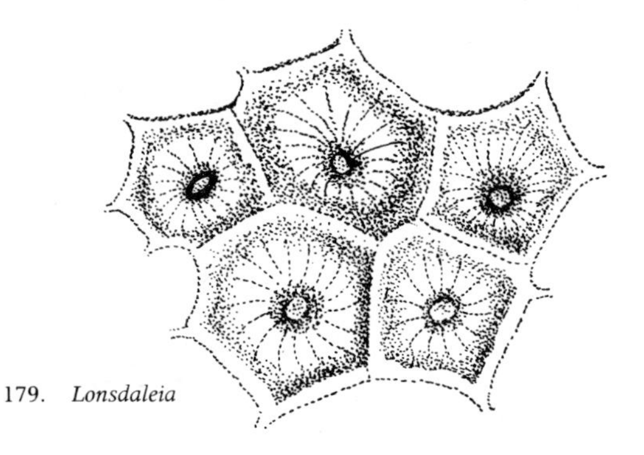

179. *Lonsdaleia*

LONSDALEIA (Fig. 179)
Carboniferous-Limestone (British Isles). A compound Coral; similar to *Lithostrotion*, with corallites circular or prismatic, but

has a wide ring of dissepiments, as a border between the septa and outer, thecal wall; has a sunken flattened central columella.

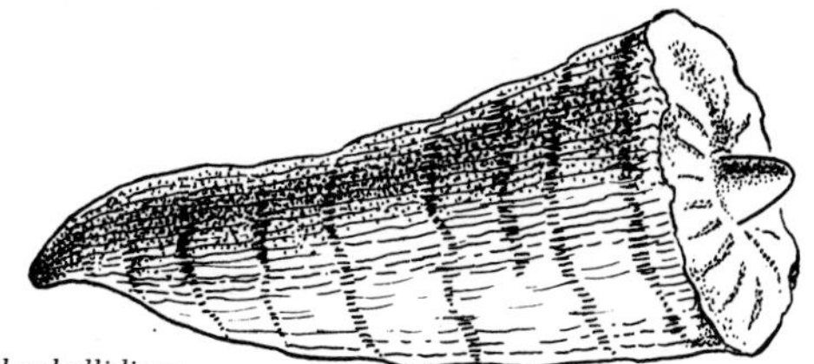

180. *Lophophyllidium*

LOPHOPHYLLIDIUM (Fig. 180)

Pennsylvanian (North America) to Permian. A solitary Coral; corallite circular, sometimes curving; the septa alternate in length as they extend towards the projecting cone-like axial rod or columella; tabulae arched; no dissepiments.

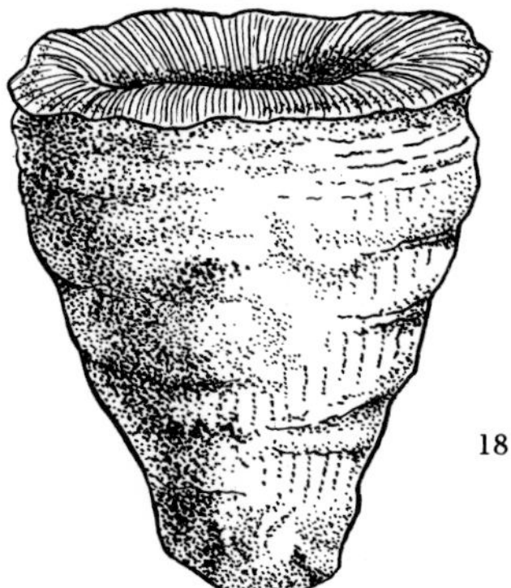

181. *Omphyma*

OMPHYMA (Fig. 181)

Silurian. A solitary Coral; corallite circular; septa prominent close to thecal wall, weak on interior; theca sunken. Not to be confused with *Caninia*.

STREPTELASMA

Ordovician to Devonian. A solitary Coral; corallite circular, sometimes curving; the numerous septa are thick bordering the thecal wall, thinner near the thecal centre, alternating in length, short and long, as they extend towards the cone-like axial rod or columella; dissepiments are weak.

ZAPHRENTIS

Carboniferous-Limestone. A solitary Coral; corallite circular, cone-shaped, sometimes curved; prominent radial septa widely spaced with fossulae or gaps; sunken theca. Not to be confused with *Streptelasma*.

HEXACORALS OR SCLERACTINIA

The *Hexacorals* or *Scleractinia* have six-fold or sextant septa, the septa frequently continuing down the thecal wall exterior. Some of the *Hexacorals* are similar to *Tabulate* and *Rugose Corals*, others having their corallites forming lines. The first *Hexacorals* occur in Triassic rocks and were particularly abundant in the Jurassic and Cretaceous Periods, but at the beginning of the Tertiary Era a large number of these became extinct, perhaps because of a change in climate. However, these were followed by Corals which continued until the present time and are the second sub-class, with the *Octocorals*, occuring as living warm-water reef-building Corals.

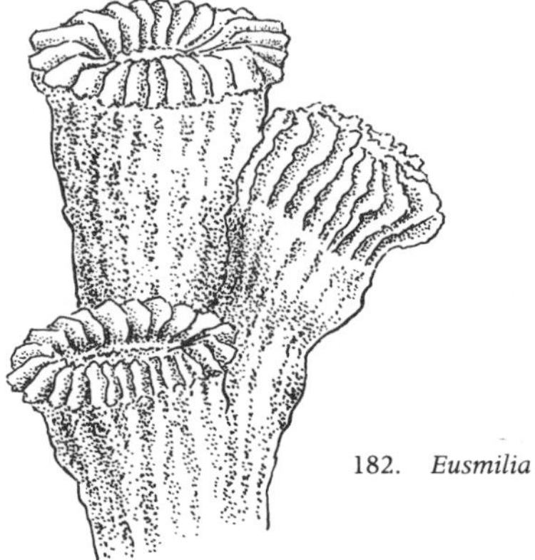

182. *Eusmilia*

EUSMILIA (Fig. 182)

Oligocene to Recent. A branched, compound Coral; the prominent septa continue down the thecal walls; no columella.

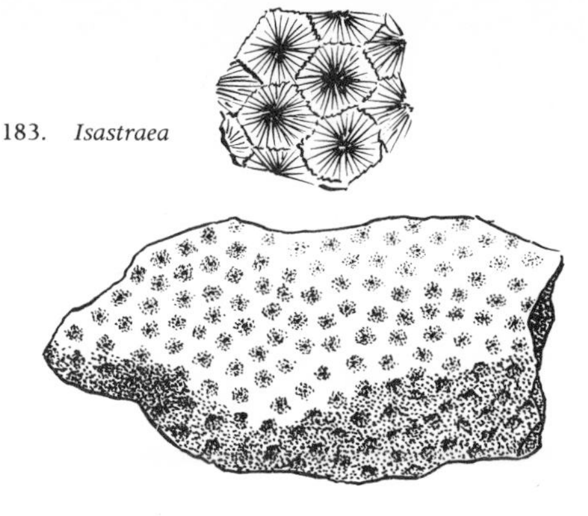

183. *Isastraea*

ISASTRAEA (Fig. 183)
Jurassic. A compound Coral; corallites approximately circular or six-sided; closely sited and fused but separated by raised walls to form a honeycomb-patterned surface; radial septa prominent.

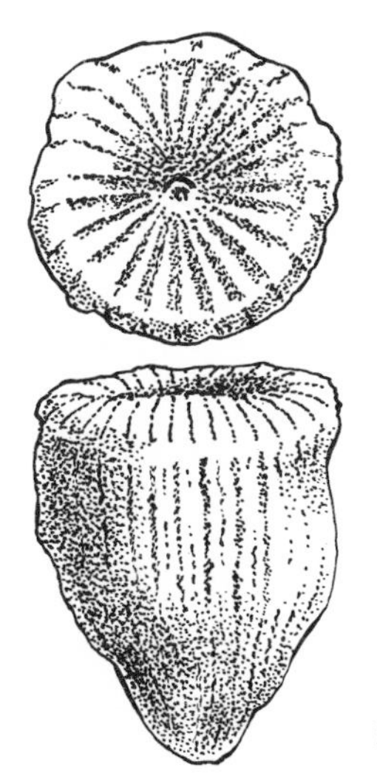

184. *Montlivaltia*

MONTLIVALTIA (Fig. 184)
Triassic, Jurassic-Inferior Oolite. A solitary Coral; corallite circular, though may be irregular cone-shaped; may be curved

or distorted; numerous long and short, radial septa, prominent; extending around upper, outer margin of thecal wall.

PARASMILIA

Cretaceous-Chalk. A solitary Coral; corallite circular, cone-shaped; slightly curved; radial septa prominent and continue down entire exterior of thecal wall.

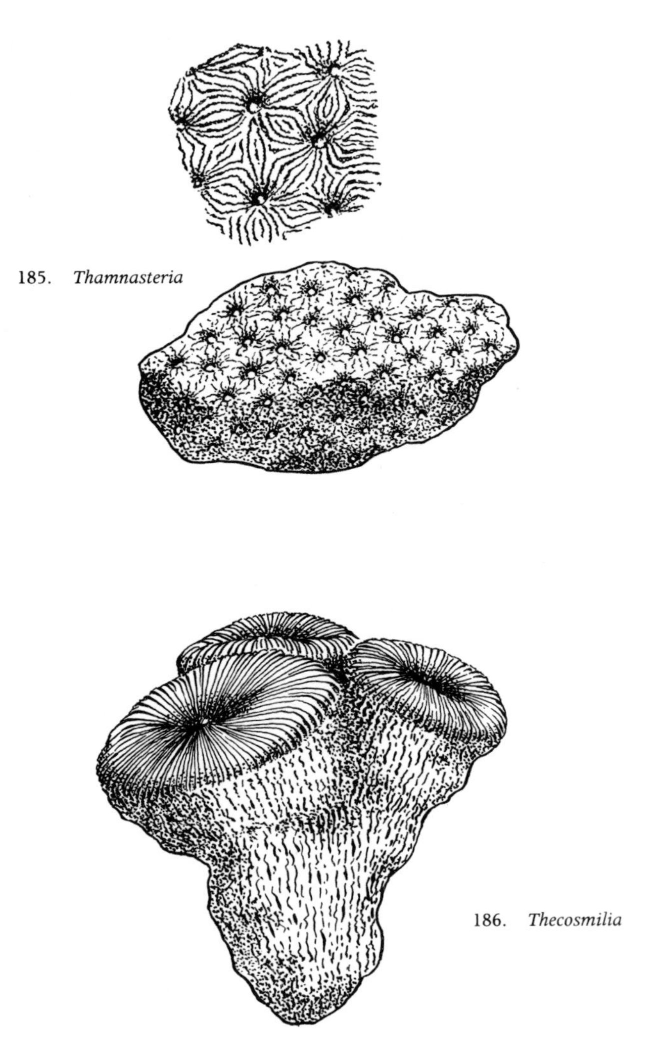

185. *Thamnasteria*

186. *Thecosmilia*

THAMNASTERIA (Fig. 185)
Triassic, Jurassic. A compound Coral; corallites closely sited and fused; thin walls indistinct or absent; septa prominent and join the adjacent corallites' septa, to give the Coral a flower-like surface pattern of beauty; has a minute columella.

THECOSMILIA (Fig. 186)
Jurassic. A branched, compound Coral; the few corallites are circular or distorted; the prominent radial septa extend around upper outer margin of thecal wall.

PORIFERA

The *Porifera*, 'pore-bearers', or Sponges, are simple in construction and, although marine animals, are plant-like in appearance, still thriving in various parts of the world. They, excepting the *Demospongea*, have an internal skeleton of fine, needle-like, calcareous or siliceous spicules that support the body. The *Porifera* is divided into three classes due to their spicule skeleton–the *Calcispongea*, having calcareous spicules; the *Hyalospongea*, having siliceous spicules; and the *Demospongea*, having siliceous spicules and also including the 'bath Sponges' that only comprise the organic material spongin. The spicules vary widely in shape and are visible under low magnification (see Fig. 187). Fossil Sponge spicules occur from Pre-Cambrian to Recent, the fossil Sponges being particularly common in the Upper Jurassic and Lower Cretaceous, also in Carboniferous Limestone. Those Sponges which had a framework of spicules to form a strong supporting skeleton are found as fairly large, recognisable fossils; but in those where the spicules were separate or non-interlocking, when the Sponge died and decomposed the individual spicules only became loose and survived. These often occur as common micro-fossils.

ASTRAEOSPONGIA
Silurian. Sponge saucer-shaped; star-like spicules prominent. Up to 2 inches in diameter.

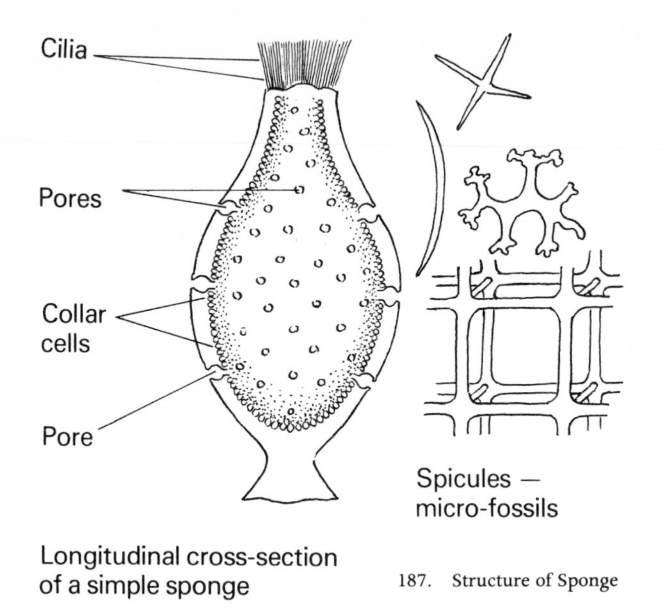

187. Structure of Sponge

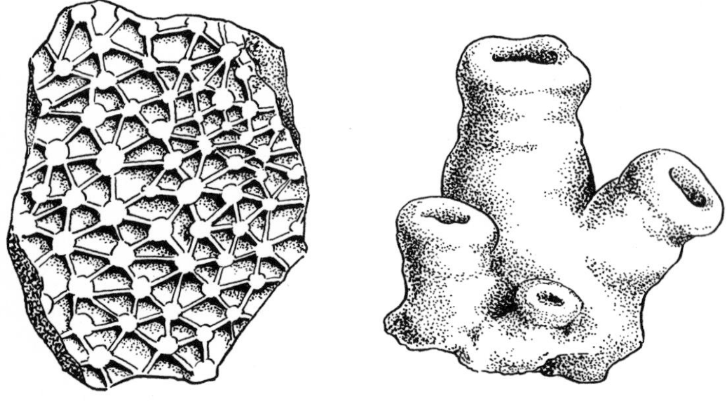

188. *Cliona*

189. *Corynella*

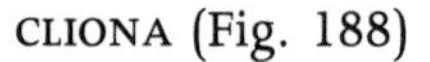

CLIONA (Fig. 188)

Upper Cretaceous-Upper Chalk. A boring Sponge which bored into the limy shells of *Lamellibranchs*. When examining fossil examples of the latter always look for *Cliona's* holes. The borings of *Cliona* are also to be found cast in flint, forming a yellow, lumpy network. It is still found today as *Cliona celata*, boring its round holes into shells, especially Oysters, and limestone rock.

CORYNELLA (Fig. 189)
Cretaceous-Lower Greensand. Sponge colonial, rounded, with irregular, perforated walls. Up to 4 inches in length.

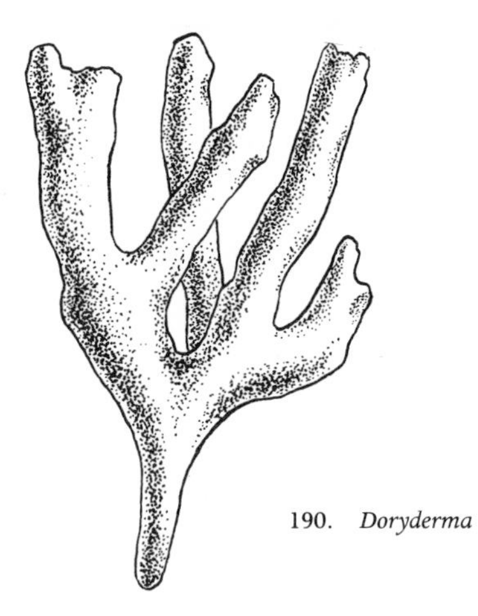

190. *Doryderma*

DORYDERMA (Fig. 190)
Upper Cretaceous-Upper Greensand, Chalk. Sponge much branched; walls with tiny perforations. Up to 5 inches in length.

HYDNOCERAS
Devonian, Mississippian (North America). Sponge octagonal; conical, irregular walls. Up to 5 inches in length.

RAPHIDONEMA (Fig. 191)
Lower Cretaceous-Lower Greensand. Sponge rounded, cup-shaped or conical, sometimes with overhanging rim, interior deep hollow; thick walls finely perforated. Up to 3 inches in height.

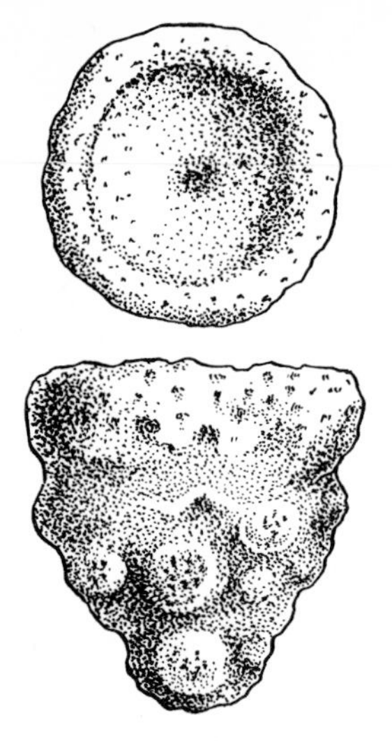

191. *Raphidonema*

VENTRICULITES (Fig. 192)
Cretaceous-Upper Chalk. Sponge conical; irregular walls perforated. Up to 4 inches in length.

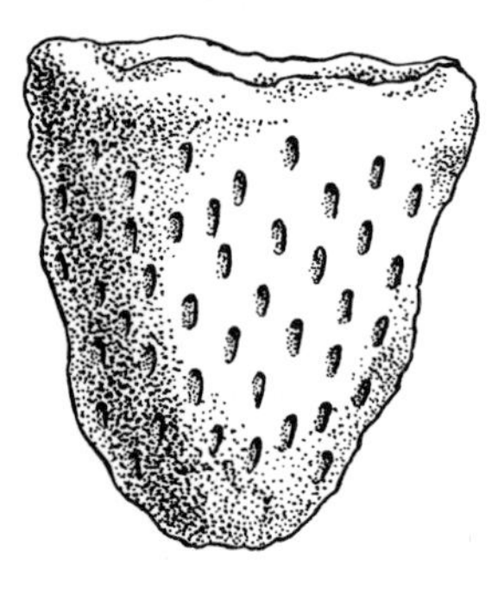

192. *Ventriculites*

SIPHONIA
Cretaceous-Upper Chalk. Sponge conical; irregular walls perforated. Up to 4 inches in length.

PROTOZOA

The *Protozoa*, 'first animals', are tiny, simple, single-celled organisms, of which there are approximately 30,000 different kinds in the sea, lakes, ponds, brackish water, and soil today. The majority are microscopic, only a few species being visible to the naked eye. They are sub-divided into four classes–*Flagellata* or *Mastigophora*; *Sarcodina* or *Rhizopoda, Sporozoa,* and *Ciliata*. The *Flagellata* and *Sarcodina* are of interest to fossil collectors, but the latter two classes do not occur as fossils.

The *Flagellata,* as their name indicates, have a whip-like thread or cord or 'flagellum', which they use to pull themselves along in water. These, however, are so minute that only the specialist micro-fossil collector with a powerful microscope can undertake their study and extraction from the sedimentary rocks.

The *Sarcodina* is more important to fossil-collectors, as it contains the two sub-classes *Foraminifera* and *Radiolaria*. *Foraminifera* occur in vast numbers, chiefly in sea water, and have a limy or chitinous skeleton test or shell, or may use 'foreign' particle-Sponge spicules, sand grains, dead *Foraminifera* shells, etc.–cemented together. As they die their tiny shells or tests accumulate on the ocean floor to form a thick deposit of ooze. They occur as fossils from the Ordovician to Recent, but were not commonplace until the Carboniferous and are particularly common in Jurassic and Cretaceous rocks, especially limestones and chalk. The majority require a hand lens or microscope for identification and close examination, although a few, such as the bulbous, cigar-shaped *Fusulinids,* Upper Carboniferous (British Isles), Pennsylvanian (North America) and Permian rocks, being up to 2 inches in length, and *Orbitolina,* Cretaceous, up to 2 inches in diameter, are visible to the naked eye. Some of the *Foraminifera* examples, indicating their diverse and variable shapes, are shown in Fig. 193.

Radiolaria also occur in vast numbers in sea water, mainly close to the surface. They have a skeleton shell of silica or strontium sulphate. As they die they, too, sink to form a deposit of ooze on the ocean floor. They occur as fossils from the Cambrian to Recent, but particularly in Ordovician and Lower

Carboniferous rocks, but are minute and must be examined by microscope. Several *Radiolaria* examples, indicating their equally diverse and variable shapes, are shown in Fig. 194.

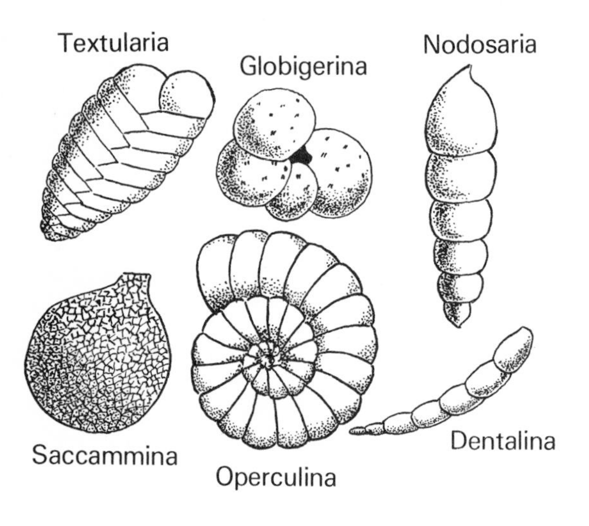

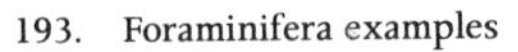
193. Foraminifera examples

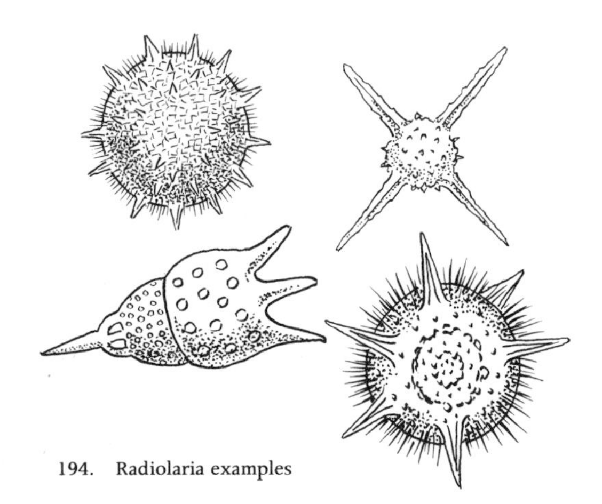
194. Radiolaria examples

PLANT FOSSILS

The fossils of plants are less often discovered than other organisms and in some instances it is a specialist task to identify them, relating their shape and construction with those of present-day species. One of the best sources of fossil plants is that of the waste tips of shale around coal mines. The fossil may sometimes be converted to coal or carbonized and have a resemblance to a 'fern leaf', but close examination may reveal whether it is a true fern, with spore cases, or a seed bearer. It may be necessary to slice the specimen and examine it under a microscope to ascertain this fact. Unfortunately for the fossil-collector, very often after the plant died it disintegrated and the separate parts became fossilized. Also during excavation and dumping the fossil may have been broken into fragments. This means that in some examples, unless the parts are found close together, it is difficult to know whether a leaf was attached to a twig fragment or a root applied to a trunk section. But where a fossil leaf is found joined to a stem it may be possible to deduce that a piece of fossil root found joined to a similar stem are all from the same plant species even if not from one plant. Fossil plant life, as Algae, occurs first in Pre-Cambrian rocks.

The fossil Plants are divided into several classes–the *Unicellular*; the *Diatoms*; the *Non-Vascular Plants-Thallophyta*; the *Multicellular-Bryophyta*; the *Vascular Plants-Pteridophytes* and *Spermaphytes*.

(Fig. 195) **Unicellular–Diatoms**

The *Diatoms* exist today as microscopic Algae, organisms that are free-floating in the sea and fresh water and thus are similar to animals, having siliceous skeletons, but in other ways are plant-like. About 10,000 living species have been recorded and some of them are identical to Cretaceous fossil *Diatoms*, but they occur from the Jurassic to Recent. *Diatomaceous earth* is the accumulated sediment or deposit formed by millions of *Diatom* skeletons, and is up to three thousand feet thick.

Non-Vascular Plants–Thallophyta

The *Thallophyta* exist today as Algae and Fungi, without roots, stems or leaves. The Algae, with chlorophyll, create their own food, but the Fungi and Slime Moulds, without chlorophyll, are parasitic. Fossil Algae occur from the Pre-Cambrian, in rocks in U.S.A., Canada, South Africa, Rhodesia, to Recent, but particularly in the Jurassic and Carboniferous Limestones. Some Algae may occur as laminated structures, while others, particularly those with soft tissues which occurred in later Periods, may not be easily recognised as such when fossilized. Fossil Algae of the Tertiary to Recent are fairly common, but fossil Algae of earlier Periods are rare. Fossil Fungi remains are extremely rare.

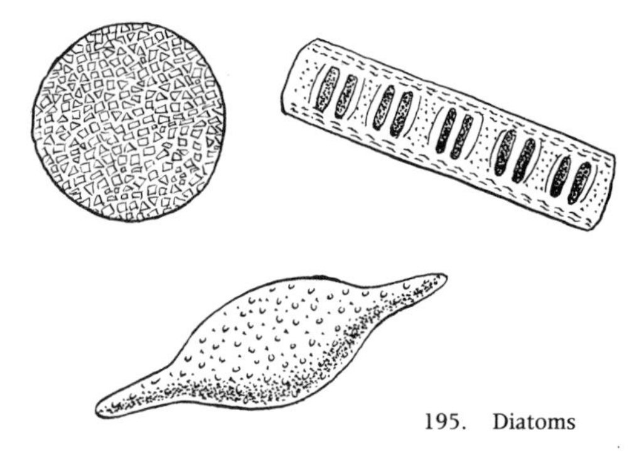

195. Diatoms

Multicellular-Non-Vascular Plants-Bryophyta

The *Bryophyta* also exist today, as the Liveworts and Mosses, being commonplace in suitable damp habitats, but their simple structure was usually too frail for fossilization and so they are rare as fossils, although they do occur from the Carboniferous (British Isles), Pennsylvanian (North America) to Recent Period.

Vascular Plants

Plants in this group *usually* have true roots, for obtaining food salts and water; a stem or stems; and leaves, with central strands of specialised conducting tissues and cells, for the water and salts progression through the plant and leaf transpiration. They are divided into the *Pteridophytes* and *Spermaphyta*.

PTERIDOPHYTES

The *Pteridophytes* reproduce by spores. The group is sub-divided into four classes, the *Psilophytales, Lycopodiales, Articulatales* and *Filicales.* In North America these are known as the *Psilopsids, Lycopods, Sphenopsids* and *Ferns.*

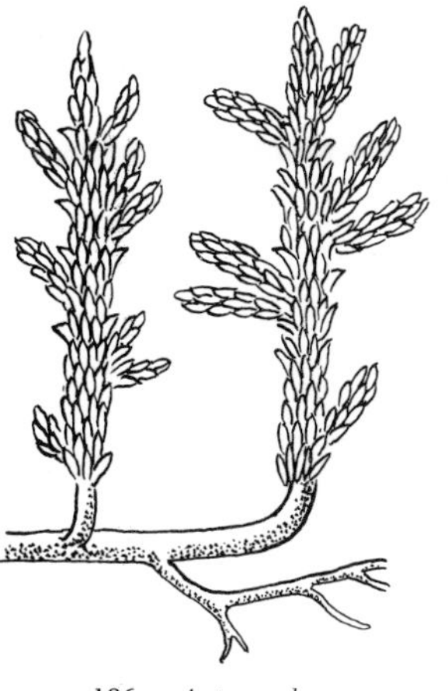
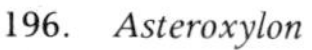
196. *Asteroxylon*

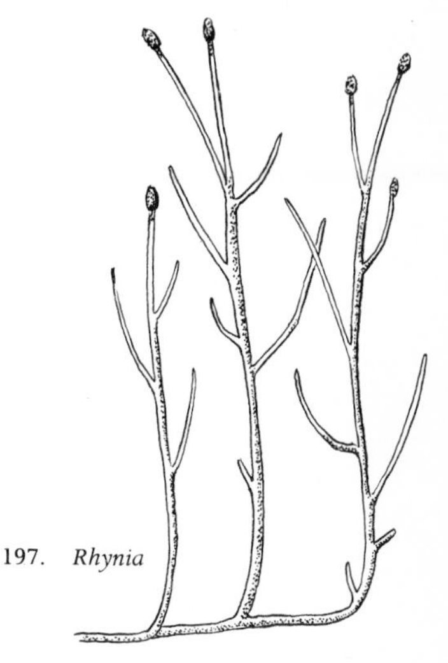
197. *Rhynia*

(Fig. 196) (Fig. 197)

PSILOPHYTALES (British Isles), *PSILOPSIDS* (North America)

These are very simple, primitive land plants with (or without) small leaves like scales; an underground rhizome and rudimentary stems and roots, the latter also being in an early stage of development. The oldest-known land plant, from the Silurian, found in Australia, is a *Psilophytale.* They were most common in the Devonian Period. Two examples are–*Asteroxylon,* Devonian, with an underground rhizome and upright stems, $\frac{1}{2}$ inch thick, bearing branches and tiny leaves. *Rhynia,* Devonian, with an underground rhizome and upright stems, branched but naked, except for spore cases or sporangia at the tips. Up to 18 inches in height.

LYCOPODIALES (British Isles), *LYCOPODS* (North America)

These exist today as the Clubmosses. Because they occur now as small plants, in some cases an inch or two in height, it is difficult to realise that in the Carboniferous (British Isles), Pennsylvanian (North America), forests they reached over 100 feet tall, with a bark-covered trunk, and numerous branches and twigs bearing simple, lance-shaped, spiky leaves and cones filled with spores. Fossil *Lycopodiales* occur from the Devonian to Recent.

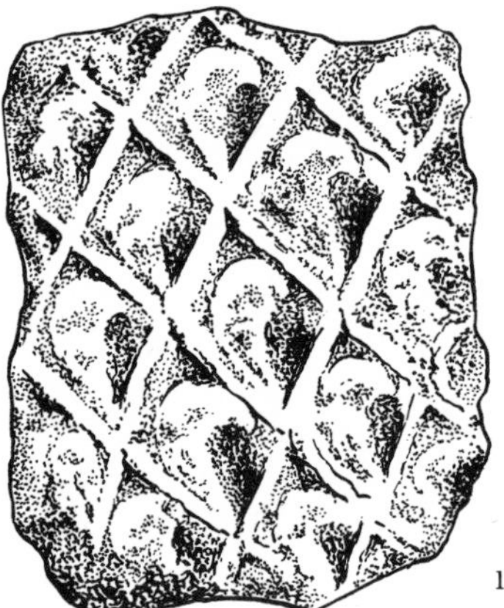

198. *Lepidodendron*

LEPIDODENDRON (Fig. 198)
Carboniferous-Coal Measures (British Isles), Pennsylvanian (North America). Tree-like, with branches, twigs and narrow leaves; trunk-bark and branches have diamond-shaped leaf scars spirally arranged. Although it reached 100 feet in height, fossils of it usually comprise fragments, especially the bark pattern.

SIGILLARIA (Fig. 199)
Carboniferous (British Isles), Pennsylvanian (North America). Tree-like, with branches and simple narrow leaves; trunk-bark and branches have vertical rows of leaf scars separated by clear grooves. Although it reached 100 feet in height, fossils of it usually comprise fragments, especially the bark pattern.

STIGMARIA (Fig. 200)
Carboniferous-Coal Measures (British Isles), Pennsylvanian

(North America). Fossil rootstock of *Lycopodiale* trees; the four main thick roots comprising the rootstock had numerous small rootlets, which became detached leaving on the pitted surface irregular spirals of scars or 'stigmata', hence their name. Complete *Stigmaria,* up to 40 feet in length, have been found, but fragments with shorter scars are more usual. These occur in the vicinity of *Fossil Forests* or the rootlets may still be attached to the rootstock of the fossil tree stumps.

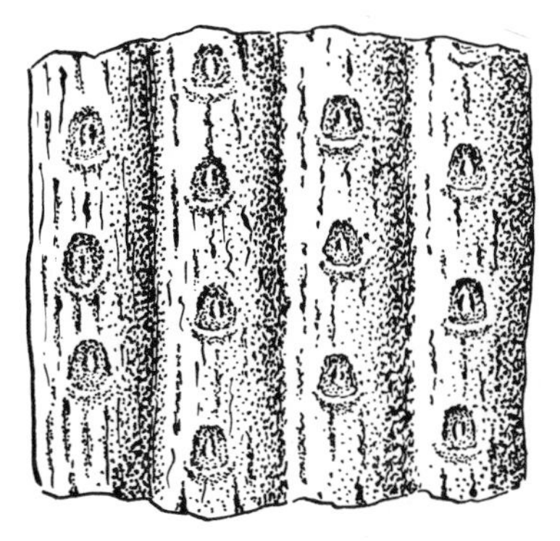

199. *Sigillaria*

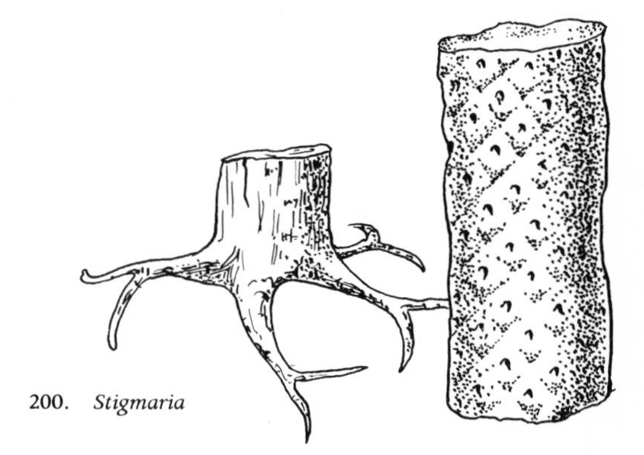

200. *Stigmaria*

ARTICULATALES (British Isles), *SPHENOPSIDS* (North America)

These have also survived until the present time as the Horse-tails and Scouring Rushes, growing from a creeping, underground rootstock or rhizome from which periodically arise erect, jointed, vertical ridged and grooved, tubular, silica-coated stems bearing symmetrical whorls of jointed, solid leaf-like branches. At the summit of the fruiting stem there is a scaled cone containing the spores. Horse-tails grow in wet marshes and waste ground usually to 5 feet in height. Like the Clubmosses it is difficult to realise that in the Carboniferous Coal Measure (British Isles), Pennsylvanian (North America) swamps their ancestral forms reached up to 60 feet in height. They occur as fossils from the Devonian to Recent.

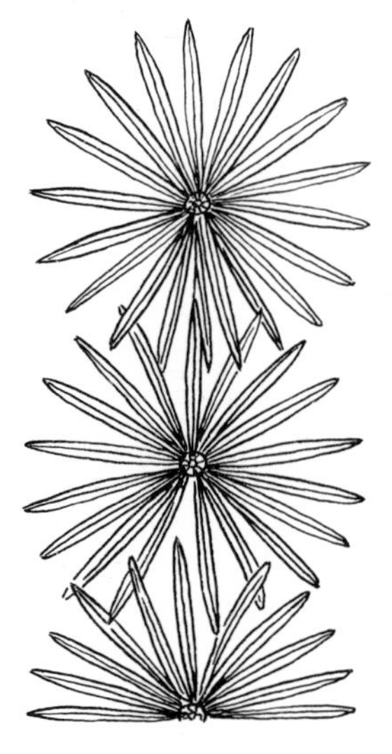

201. *Annularia*

ANNULARIA (Fig. 201)

Upper Carboniferous (British Isles). Foliage of a Horse-tail in whorls; leaves narrow, lance-shaped.

CALAMITES (Fig. 202)

Upper Carboniferous-Coal Measures (British Isles), Mississippian (North America) to Permian. A Scouring Rush, close relative of the Horse-tails. Hollow stem vertically ribbed, with nodes or joints, sites of the leaf-whorls. Tree-size, up to 40 feet in height, but as a fossil is usually a fragment section of the stem.

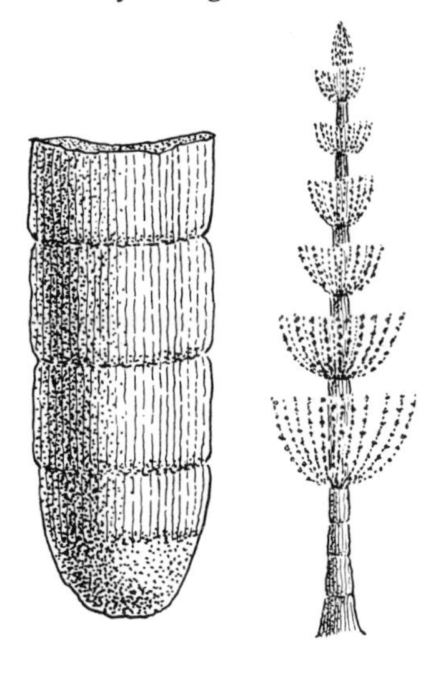

202. *Calamites*

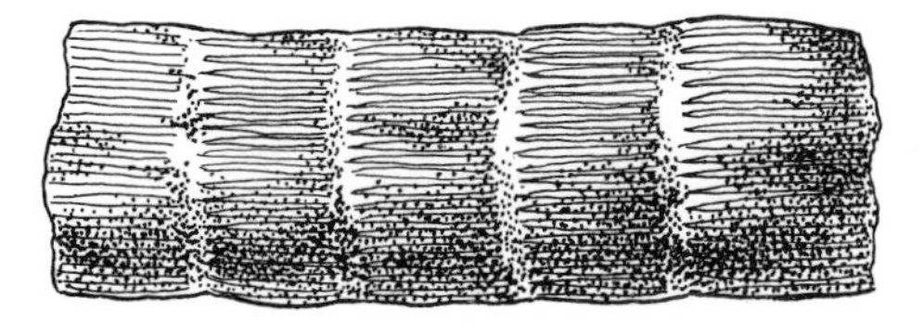

203. *Equisetites*

EQUISETITES (Fig. 203)
Cretaceous. A Horse-tail; ancestor of the present-day *Equisetum*. Vertical ribbed stem jointed with sites of leaf whorls. Fossils usually occur as fragments of stem, sometimes still in an upright position as if growing.

SPHENOPHYLLUM (Fig. 204)
Devonian to Triassic. A climbing plant with narrow, ribbed stems; leaves wedge-shaped, in close whorls of six leaves or a multiple of three.

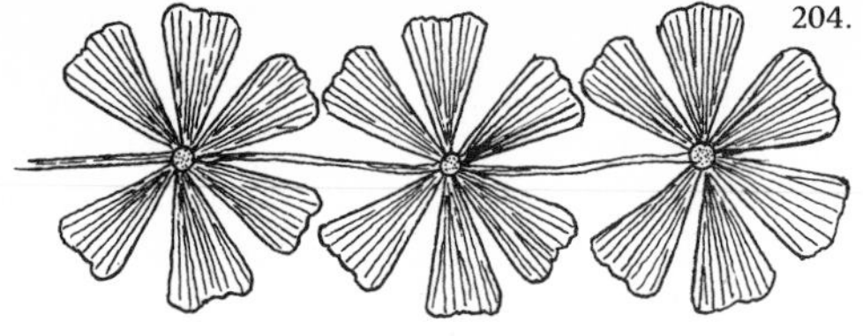
204. *Sphenophyllum*

FILICALES

These survive today as the Ferns. Their large, complex leaf-shape is familiar, with the sporangia or spore cases on the undersides. They occur as fossils from the Devonian to Recent, but became common in the Pennsylvanian (North America), Carboniferous (British Isles). Care should be taken when identifying fossils which appear to be Ferns in order to avoid confusion between the true Ferns and the extinct Seed Ferns, described in *Gymnosperms*.

SPERMAPHYTA

The *Spermaphyta* reproduce by seeds. The group is sub-divided into the *Gymnosperms* and *Angiosperms*.

GYMNOSPERMS

The *Gymnosperms*, 'naked seeds', bear no flowers and the seeds are not entirely enclosed in an ovary. They are divided into four classes–the *Pteridospermeae, Cycadophyta, Ginkgoales* and *Coniferales*. Some species are extinct, others survive to the present day.

PTERIDOSPERMEAE

The *Pteridosperms*, now extinct, were an important feature of the undergrowth of the Coal Measure forests and included the seed-bearing Ferns. These developed seeds upon their leaves, not in the shape of a cone as is usual among *Gymnosperms*. They occur as fossils from the Devonian to Jurassic.

ALETHOPTERIS (Fig. 205)

Carboniferous (British Isles), Pennsylvanian (North America). A seed Fern with fronds of numerous long, blade-like leaflets or pinnules; the rib-veins are usually clearly defined.

Index